APPLIED DIGITAL SIGNAL PROCESSING AND APPLICATIONS

OTHMAN OMRAN KHALIFA

PARTRIDGE

Copyright © 2021 by Othman Omran Khalifa.

ISBN:　　　Softcover　　　978-1-5437-6629-5
　　　　　　eBook　　　　978-1-5437-6630-1

All rights reserved. No part of this book may be used or reproduced by any means, graphic, electronic, or mechanical, including photocopying, recording, taping or by any information storage retrieval system without the written permission of the author except in the case of brief quotations embodied in critical articles and reviews.

Because of the dynamic nature of the Internet, any web addresses or links contained in this book may have changed since publication and may no longer be valid. The views expressed in this work are solely those of the author and do not necessarily reflect the views of the publisher, and the publisher hereby disclaims any responsibility for them.

Print information available on the last page.

To order additional copies of this book, contact
Toll Free +65 3165 7531 (Singapore)
Toll Free +60 3 3099 4412 (Malaysia)
orders.singapore@partridgepublishing.com

www.partridgepublishing.com/singapore

Contents

Preface xi
Dedication xiii

Chapter 1 Introduction to Signals 1
  1.1 Introduction 1
  1.2 Signal Classification 3
      1.2.1 Continuity of the independent and
           dependent variables 3
      1.2.2 Predictability of the dependent variables
           with respect to the independent variable. 5
      1.2.3 Dimensionality of Signals 7
      1.2.4 Periodic vs. Aperiodic Signals 8
      1.2.5 Causal vs. Anticausal Signals 9
      1.2.6 Even vs. Odd Signals 10
      1.2.7 Energy vs. Power Signals 11
  1.3 Elementary Signals 12
      1.3.1 Unit Impulse Function 12
      1.3.2 Unit Step Function 13
      1.3.3 Rectangular Pulse Function 14
      1.3.4 Signum function 14
      1.3.5 Ramp function 15
      1.3.6 Sinc function 16
      1.3.7 Exponential Function 16

Chapter 2 Introduction to Systems 23
  2.1 Introduction 23
  2.2 Classification of Systems 26
      2.2.1 Linear and non-linear systems 27
      2.2.2 Time-varying and time-invariant systems 29
      2.2.3 Static and Dynamic Systems 31
      2.2.4 Invertible and non-invertible systems 33
      2.2.5 Causal and non-causal systems 34
      2.2.6 Stable and unstable systems 36
  2.3 Impulse Response and Convolution 38

Chapter 3   Sampling, Quantization and Reconstruction 49
   3.1 Introduction.. 49
   3.2 Signal Sampling... 51
   3.3 Interpolation.. 53
   3.4 The Sampling Theorem 54
   3.5 Aliasing.. 56
   3.6 Antialiasing Prefilters 57
   3.7 Types of Sampling 58
      3.7.1 Impulse (Ideal) Sampling 59
      3.7.2 Natural Sampling... 60
      3.7.3 Sample-and-Hold (Flat Top) Sampling 62
   3.8 Quantization.. 65
      3.8.1 Quantization Error... 66
   3.9 Ideal Reconstruction 68
   3.10 Signal Reconstruction.. 71

Chapter 4   Discrete-Time Signals and Systems 77
   4.1 Discrete-Time Signals... 77
      4.1.1 Some Elementary Sequences........................... 78
         4.1.1.1 Unit Impulse Sequence 79
         4.1.1.2 Unit Step Sequence 79
         4.1.1.3 The unit ramp signal........................... 79
         4.1.1.4 Sinusoidal Sequences 80
         2.1.1.5 Complex Exponential Sequences........... 81
         4.1.1.6 Random Sequences 81
      4.2 Types of Sequences.. 83
         4.2.1 Real vs. Complex Signals................................ 83
         4.2.2 Finite vs. Infinite Length 83
         4.2.3 Causal vs. Anti-casual Signals........................ 83
         4.2.4 Energy and Power Signals.............................. 86
      4.3 Some Basic Operations on Sequences...................... 90
      4.4 Discrete-time Systems 92
         4.4.1 Classification of Systems 92
         4.4.2 Linear Shift-Invariant Systems........................ 98
         4.4.3 Linear Convolution... 99
         4.4.4 Properties of Linear Convolution 102

   4.4.4.1 Condition for Stability............................102
   4.4.4.2 Condition for Causality.........................103

Chapter 5 Z-transform and applications 108
 5.1 Introduction...108
 5.2 Unilateral Z-transform 111
 5.3 Bilateral Z-transform 117
 5.4 Poles and Zeros in the Z-Plane 119
 5.5 Properties of the z transform.......................................121
 5.6 Region of Convergence for the Z-Transform 125
  5.6.1 Properties of the Region of Convergence........125
 5.7 Inverse z-Transform...131
  5.7.1 Power Series Method..132
  5.7.2 Partial Fraction Expansion 133
  5.7.3 Contour integration...139
 5.8 Transfer Function in the Z-domain 140
 5.9 Application to signal processing 141
  5.9.1 Solution of Difference Equations Using
   the z-Transform 141
  5.9.2 Analysis of Linear Discrete Systems...............147

Chapter 6 Frequency Analysis of Discrete Signals
    and Systems... 152
 6.1 Introduction...152
 6.2 Frequency analysis of a Continuous Time signal 153
  6.2.1 Fourier Series for Continuous-Time
   Periodic Signals..154
 6.3 Frequency Analysis of Discrete-Time Signals..........160
  6.3.1 Fourier Series for Discrete-Time
   Periodic Signals..161
  6.3.2 Fourier Transform of Discrete-Time
   Aperiodic Signals 166
 6.4 Frequency Domain Representation of
  Discrete-time LTI Systems 169
  6.4.1 Steady State Response of LTI
   Discrete-time Systems 172

6.5 Frequency Response of Systems 177
6.6 Convolution via the Frequency Domain 180

Chapter 7   Discrete Fourier Transform 185
  7.1 Introduction 185
  7.2 DFT as matrix multiplication 192
  7.3 Properties of the DFT 195
    7.3.1 Periodicity 195
    7.3.2 Orthogonality 196
    7.3.3 Linearity 196
    7.3.4 Hermitian symmetry 196
    7.3.5 Time shifting 197
    7.3.6 Circular convolution 197
    7.3.7 Parseval's theorem 198
  7.4 Computational complexity 199
  7.5 Fast Fourier Transform (FFT) 201
    7.5.1 Derivation of the FFT 202

Chapter 8   Design of Digital Filters 213
  8.1 Introduction 213
    8.1.1. Finite Impulse Response 217
    8.1.2 Infinite Impulse Response 218
    8.1.3 Filter Specification Requirements 218
  8.2 FIR Digital Filters 219
    8.2.1 Design of FIR Digital Filters using Impulse Response Truncation (IRT) 225
    8.2.2 Design of FIR filters using windowing technique. 227
    8.2.3 Design of FIR filters by frequency sampling 235
  8.3 Design of IIR Filters 239
    5.3.1 IIR Filter Basics 240
    8.3.2 Bilinear transformation method 242
    8.3.3 Analog Filter using lowpass prototype Transformation 243
    8.3.4 Bilinear Transformation and Frequency Warping 248
    8.3.5 Bilinear Transformation Design Procedure 255

8.4.6 Impulse Invariant Design Method 260

Chapter 9   Wavelet Transform 271
9.1 Introduction 271
9.2 Continuous Wavelet Transform 272
9.3 Time-Frequency Resolution 277
9.4 Wavelet Series 279
   9.4.1 Dyadic Sampling 279
9.5 Discrete Wavelet Transform (DWT) 282
   9.5.1 Multiresolution Analysis 283
   8.5.2 Wavelet Analysis by Multirate Filtering 288
   8.5.3 Wavelet Synthesis by Multirate Filtering 290
9.6 Discrete Wavelet Transform for denoising data ... 291
9.7 Signal denoising for IoT networks 293
9.8 Multiresolution Signal Analysis 294
9.9 Multiresolution Wavelet Decomposition of
     Transient Signal 295
9.10 Signal Detection 297

Chapter 10 Adaptive Signal Processing 302
10.1 Introduction 302
10.2 Adaptive Noise Cancellation 304
10.3 Adaptive Filtering Algorithms 305
   10.3.1 Least Mean Square (LMS) Algorithm 309
   10.3.2 The Recursive Least Squares (RLS)
       Algorithm 311
   10.3.3 Wiener Filtering 313
      10.3.3.1 Adaptive Wiener Filter 315
10.4 Applications of Adaptive Filters 316
   10.4.1 System Identification 316
   10.4.2 Channel Identification 318
   10.4.3 Plant Identification 319
   10.4.4 Echo Cancellation for Long-Distance
       Transmission 319
   10.4.5 Acoustic Echo Cancellation 320
   10.4.6 Adaptive Noise Cancelling 321
10.5 Inverse Modeling 322

    10.5.1 Channel Equalization 323
    10.5.2 Inverse Plant Modeling 323
    18.5.3 Linear Prediction 324
        10.5.3.1 Linear Predictive Coding 325
    10.5.4 Adaptive Line Enhancement 325
    10.6 Adaptive Noise Reduction 326

References.. 333

Preface

Due to the rapid development of technologies, digital information playing a key role in our daily life. In the past signal processing appeared in various concepts in more traditional courses where the analog and discrete components were used to achieve the various objectives. However, in the 21th century, with the rapid growth of computing power in terms of speed and memory capacity and the intervention of artificial intelligent, machine /deep learning algorithms introduces a tremendous growth in signal processing applications. Therefore, digital signal processing has become such a critical component in contemporary science and technology that many tasks would not be attempted without it. It is a truly interdisciplinary subject that draws from synergistic developments involving many disciplines. The developers should be able to solve problems with an innovation, creativity and active initiators of novel ideas. However, the learning and teaching has been changed from conventional and tradition education to outcome based education. Therefore, this book prepared on a Problem-based approach and outcome based education strategies. Where the problems incorporate most of the basic principles and proceeds towards implementation of more complex algorithms. Students required to formulate in a way to achieve a well-defined goals under the guidance of their instructor.

This book follows a holistic approach and presents discrete-time processing as a seamless continuation of continuous-time signals and systems, beginning with a review of continuous-time signals and systems, frequency response, and filtering. The synergistic combination of continuous-time and discrete-time perspectives leads to a deeper appreciation and understanding of DSP concepts and practices.

This book is organized in Ten chapters as follows: Chapter One, introduces the basic terminology of signals in digital signal processing. Classification of signals as well as the elementary signal are explained in detail. Chapter Two describes the concept of systems and characterize and analyze the properties of Discrete systems. Chapter Three covers the sampling process, Quantization, coding and reconstruction of signals. Chapter Four introduces the properties of discrete signals and systems. Chapter Five introduces the z-transform and difference equations and its applications. Chapter Six explains the frequency analysis of Discrete Signals and Systems, Frequency Response of Systems and convolution via frequency domain. Chapter Seven devoted for Discrete Fourier transform. Chapter Eight deals with various methods used in Digital filters design. Chapter Nine introduces the wavelet transforms, Multiresolution Analysis and some applications of discrete wavelet transform. Chapter Ten deals with adaptive signal processing and covers Wiener filter, LMS algorithms, RLS algorithms and ends with applications of adaptive filters.

Author
Othman Omran Khalifa

Dedication

To my family: the soul of my father, the lovely mother, wife and children

Chapter One

Introduction to Signals

Learning Outcomes of this Chapter

After successful completion of this chapter students will be able to:

1. understand basic terminology in digital signal processing.
2. differentiate digital signal processing and analog signal processing.
3. characterize and analyze the properties of Discrete time signals.
4. describe signals mathematically and understand how to perform mathematical operations on signals.
5. describe basic digital signal processing application areas.

1.1 Introduction

Signals are detectable quantities used to convey information about time-varying physical phenomena. Common examples of signals are human speech, temperature, pressure, and stock prices. Electrical signals, normally expressed in the form of voltage or current waveforms, are some of the easiest signals to generate and process. Mathematically, signals are modeled as functions of one or more independent variables. Examples of independent variables used to represent signals are time, frequency, or spatial coordinates. Before introducing the mathematical notation used to represent signals,

Let us consider a few physical systems associated with the generation of signals. When we want to observe the real world, we need a measuring instrument connected to an information system. A basic block diagram of such a set-up is sketched in figure 1.1. The first component is a *sensor* or *transducer* to convert the physical quantity we are interested in into an electrical signal. For instance, for sound we need a microphone to convert variations in air pressure into an electrical signal. For images we may use a video camera to obtain a video signal which represents the brightness in the image when it is scanned line by line.

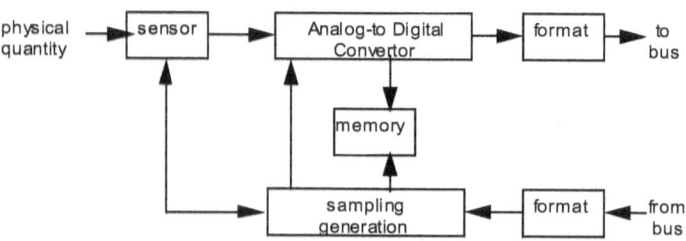

Figure 1.1 Basic model of a measuring instrument

The next block represents the conversion of the electrical signal into digital numbers. This is realized by an Analog-to-Digital Converter (ADC). The input range of the ADC is divided into a large number of intervals of equal size Δv. The successive intervals are numbered to represent the quantized input. So, when the number k is assigned to the quantized signal, the original value v was in the interval between v_k and $v_k + 1$:

$$v_k \leq v < v_k + 1$$

This process is illustrated in figure 1.2 for 8 quantization intervals. The number of quantization levels is in general a power of 2. When we have n bits available the number of quantization levels is 2n. For example, when the number of bits n = 8 there are 256 intervals, and the *resolution* is said to be 256.

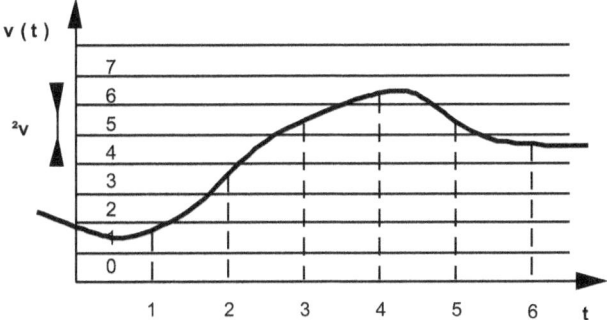

Figure 1.2 Quantization process of a 3bit ADC with 8 quantization levels. The successive quantized values of v for t= 1 through 6 are 1,3,5,6,5,4.

An important decision to be made is the number of quantization levels (so the number of bits) needed to represent the continuous signal. This is related to the noise (inaccuracy) present in the sensor signal. The inaccuracy introduced by the quantization process should be considerably smaller than the inaccuracy in the sensor signal itself. The details will be discussed in later chapters.

1.2 Signal Classification

A signal is classified into several categories depending upon the criteria used for its classification. In this section, we cover the following categories for signals:

1.2.1 Continuity of the independent and dependent variables

i. Continuous-time signal: The time variable is continuous in the range in which the signal is defined. If the signal variable is represented by *x*, time variable is *t* such a signal is denoted as x(t). However, if a signal is defined for all values of the independent variable *t*, it is called a *continuous-time* (CT) signal. Consider the signals shown in figure 1.3. Since these signals vary continuously with time *t* and have known magnitudes for all time instants, they are classified as CT signals.

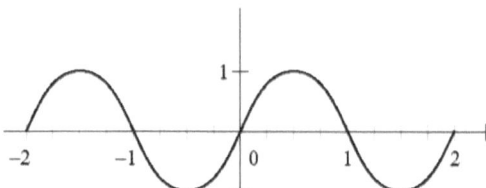

Figure 1.3 Continuous-time Signal

ii. Discrete-time signal: The time variable is discrete in the range in which the signal is defined. If the signal variable is x and the time variable has been sampled at time instances *n*, where $n = n'T$ then the signal is denoted as x(n). A discrete time signal is also referred to as a sampled signal since it is obtained by directly sampling a targeted signal. It should be noted that the amplitude of the sampled signal can take any value within a specified amplitude range, and we therefore say that the amplitude of discrete-time signal is continuous. if a signal is defined only at discrete values of time, it is called a *discrete time* (DT) signal as shown in figure 1.4. (e.g the value of a stock at the end of each month)

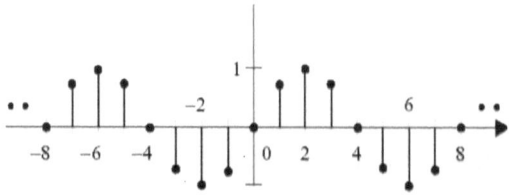

Figure 1.4 Discrete-Time Signal

A digital signal: This is a signal that is discrete in time and discrete in amplitude. It is represented in the same way as a discrete-time signal.

1.2.2 Predictability of the dependent variables with respect to the independent variable.

i. A signal is said to be deterministic if the dependent variable is predictable at any instance of the independent variable time. The signal is a signal in which each value of the signal is fixed and can be determined by a mathematical expression, rule, or table. Because of this, the future values of the signal can be calculated from past values with complete confidence.

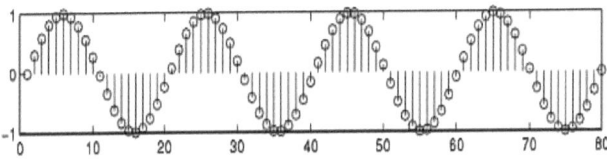

Figure 1.5 (a) Discrete-Time Signal v.s Original signal

ii. A random signal, on the hand, has an unpredictable dependent variable at any instance of the independent variable time. Such a signal can only be defined in terms of its statistical properties. a random signal has a lot of uncertainty about its behavior. The future values of a random signal cannot be accurately predicted and can usually only be guessed based on the averages of sets of signals for example Electrical noise generated in an amplifier of a radio/TV receiver.

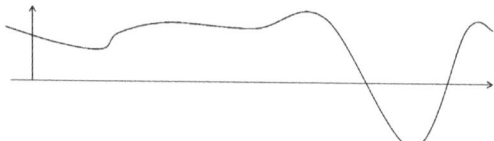

Figure 1.5 (b) Random Signal

Example 1.1

Consider the CT signal $x(t) = sin(t)$ plotted in Fig. 1.6(a) as a function of time t. Discretize the signal using a sampling interval of $T = 0.25$ sec., and sketch the waveform of the resulting DT sequence for the range $-2 \leq k \leq 9$.

Solution:

By substituting $t = kT$, the DT representation of the CT signal $x(t)$ is given
by $x[kT] = sin(k \times 0.25) = sin(0.25k)$
For $k = 0, \pm 1, \pm 2, \ldots$, the DT signal $x[k]$ has the following values:

$x[-8] = x(-8T) = sin(-2\pi) = 0, x[1] = x(T) = sin(0.25\pi) = 1/\sqrt{2}$

$x[-7] = x(-7T) = sin(-1.75\pi) = 1/\sqrt{2}, x[2] = x(2T) = sin(0.5\pi) = 1$

$x[-6] = x(-6T) = sin(-1.5\pi) = 1, x[3] = x(3T) = sin(0.75\pi) = 1/\sqrt{2}$

$x[-5] = x(-5T) = sin(-1.25\pi) = 1/\sqrt{2}, x[4] = x(4T) = sin(\pi) = 0$

$x[-4] = x(-4T) = sin(-\pi) = 0, x[5] = x(5T) = sin(1.25\pi) = -1/\sqrt{2}$

$x[-3] = x(-3T) = sin(-0.75\pi) = -1/\sqrt{}, x[6] = x(6T) = sin(1.5\pi) = -1$

$x[-2] = x(-2T) = sin(-0.5\pi) = -1, x[7] = x(7T) = sin(1.75\pi) = -1/\sqrt{2}$

$x[-1] = x(-T) = sin(-0.25\pi) = -1/\sqrt{2}, x[8] = x(8T) = sin(2\pi) = 0,$
$x[0] = x(0) = sin(0) = 0.$

Plotted as a function of k, the waveform for the DT signal $x[k]$ is shown in Fig. 1.6(b), where for reference the original CT waveform is plotted with a dotted line. We will refer to a DT plot illustrated in Fig. 1.6(b) as a *bar* or a *stem* plot to distinguish it from the CT plot of $x(t)$, which will be referred to as a *line* plot.

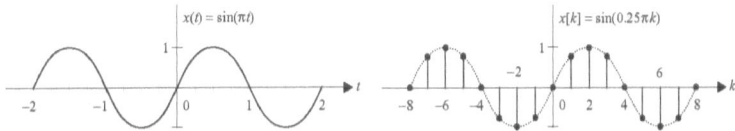

Figure 1.6. (a) CT sinusoidal signal (b) DT sinusoidal signal $x[k]$

1.2.3 Dimensionality of Signals

All the above classifications of digital signals can further be classified in terms of their Dimensionality. Here, we will only elaborate this classification using discrete-time sequences and we will leave the rest to the student.

i. A one-dimensional signal has only one-independent variable and one-dependent variable. A discrete-time signal $x(n)$ is a one-dimensional signal as it has only one-independent variable, discrete-time (n), and one-dependent variable, the amplitude of $x(n)$.

ii. A two-dimensional signal has two-independent variables and one-dependent variable. The samples n and m are taken in the spatial domain. The two-dimensional signal is discrete in the spatial domain in two-dimensions. The independent variables are n, m which define the dependent variable $x(n,m)$. A good example is a photographic image where n,m define the spatial location and $x(n,m)$ defines the grey level at the location.

iii. A three-dimensional signal has three-independent variables and one-dependent variable. A discrete-time signal $x(n,m,\tau)$ is a three-dimensional signal as it has two-independent variable in the spatial domain (n,m) and one-independent variable τ in the time domain. The three-independent variables define the one-dependent variable, the intensity of $x(n,m,\tau)$. An example of a three-dimensional signal is video signal where a signal at spatial location (n,m) is changing with respect to time τ.

1.2.4 Periodic vs. Aperiodic Signals

Periodic signals is a function of time that repeat it self with some period T to satisfies the following:

$$x(t) = x(t+T) \tag{1.1}$$

The smallest T, that satisfies this relationship is called the fundamental period.

Likewise, a DT signal $x[k]$ is said to be *periodic* if it satisfies:

$$x[n+N] \tag{1.2}$$

at all time n and for some positive constant N. The smallest positive value of N that satisfies the periodicity condition, A signal that is not periodic is called an *aperiodic* or *non-periodic* signal. Figure 1.7 shows examples of both periodic and aperiodic.

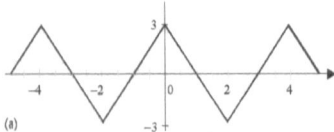

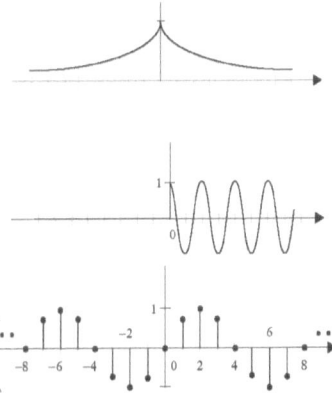

Figure 1.7. Examples of periodic and aperiodic signals.

1.2.5 Causal vs. Anticausal Signals

Causal Signals are signals that are zero for all negative time. Causality in systems makes the most sense. Causality in signals doesn't make that much sense. Causality in a system determines whether a system relies on future information of a signal $x[n+1]$. When talking about "causality" in signals, we mean whether they are zero to the left of $t = 0$ or zero to the right of $t = 0$.

A causal signal is zero for $t < 0$. However, the reason why this doesn't really make sense is that if you have a signal, the time $t = 0$ can be chosen arbitrarily.

A continuous time signal $x(t)$, is said to be casual if : $x(t) = 0$ for every $t < 0$, the signal $x(t)$ does not start before $t = 0$.

Figure 1.8. Examples of causal and non causal signals

1.2.6 Even vs. Odd Signals

An even signal is any signal f such that $x(t) = x(-t)$. Even signals can be easily spotted as they are symmetric around the vertical axis. Sin(t) is an odd signal.

An odd signal, on the other hand, is a signal f such that $x(t) = -x(-t)$. $Cos(t)$ is an even signal. Also, a signal can be even, odd or neither.

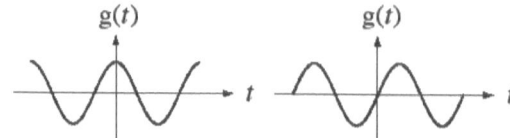

Figure 1.9. Sin(t) and Cos(t)

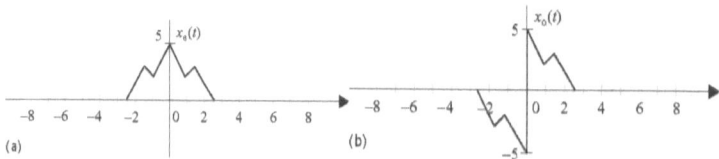

Figure 1.10: Other examples of Even and Odd signals

Using the definitions of even and odd signals, we can show that any signal can be written as a combination of an even and odd signal. That is, every signal has an odd-even decomposition.

$$x_e(t) = \frac{x(t) + x(-t)}{2}$$
$$x_o(t) = \frac{x(t) - x(-t)}{2}$$
(1.3)

1.2.7 Energy vs. Power Signals

Energy Signal is a signal with finite energy and zero power is called Energy Signal i.e. for energy signal

$$0 < E < \infty \text{ and } P = 0$$

Signal energy of a signal is defined as the *area under the square of the magnitude of the signal*. The units of signal energy depends on the unit of the signal.

Some signals have infinite signal energy. In that case it is more convenient to deal with **average signal power.**

$$E_x = \int_{-\infty}^{\infty} |x(t)|^2 \, dt \qquad (1.4)$$

For power signals

$$0 < P < \infty \text{ and } E = \infty$$

Average power of the signal is given by

$$P_x = \lim_{T \to \infty} \frac{1}{T} \int_{-T/2}^{T/2} |x(t)|^2 \, dt \qquad (1.5)$$

For a periodic signal x(*t*) the average signal power is

$$P_x = \frac{1}{T} \int_T |x(t)|^2 \, dt \qquad (1.6)$$

T is any period of the signal.
   Periodic signals are generally power signals

1.3 Elementary Signals

In this section, we define some elementary functions that will be used frequently to represent more complicated signals. Representing signals in terms of the elementary functions simplifies the analysis and design of linear systems.

1.3.1 Unit Impulse Function

The unit impulse (or Dirac delta function) is a signal that has infinite height and infinitesimal width. However, because of the way it is defined, it integrates to one. While in the engineering world, this signal is quite nice and aids in the understanding of many concepts, some mathematicians have a problem with it being called a function, since it is not defined at $t = 0$. Engineers reconcile this problem by keeping it around integrals, in order to keep it more nicely defined. The unit impulse is most commonly denoted as $\delta(t)$.

The most important property of the unit-impulse is shown in the following integral:

$$\int_{-\infty}^{\infty} \delta(t) dt = 1 \qquad (1.7)$$

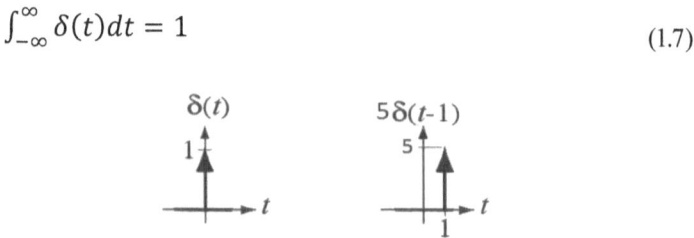

Figure 1.11: Impulse Function representation

However, the unit impulse train is a sum of infinitely uniformly-spaced impulses and is given by:

$$\delta_T(t) = \sum_{n=-\infty}^{\infty} \delta(t-nT), \quad n \text{ an integer}$$

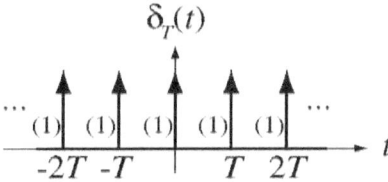

Figure 1.12: Unit impulse train

1.3.2 Unit Step Function

The waveforms for the unit step functions $u(t)$ and $u[k]$ are shown respectively, in figure 1.13. It is observed that the CT unit step function $u(t)$ is piecewise continuous with a discontinuity at $t = 0$ is defined as follows:

$$u(t) = \begin{cases} 1 & t \geq 0 \\ 0 & t < 0 \end{cases} \tag{1.8}$$

However, the DT function $u[k]$ has no such discontinuity is defined as follows:

$$u[k] = \begin{cases} 1 & k \geq 0 \\ 0 & k < 0 \end{cases}$$

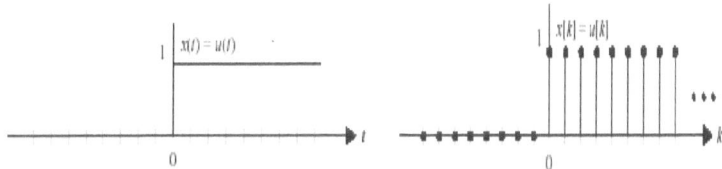

Figure 1.13 Unit step function

1.3.3 Rectangular Pulse Function

The CT rectangular pulse $rect(t/\tau)$ is defined as follows:

$$rect(\frac{t}{\tau}) = \begin{cases} 1 & |t| \geq \tau/2 \\ 0 & |t| < \tau/2 \end{cases} \quad (1.9)$$

While the DT rectangular pulse $rect(k/(2N + 1))$ is defined as follows:

$$rect(\frac{k}{2N+1}) = \begin{cases} 1 & |k| \geq N \\ 0 & |k| < N \end{cases}$$

The waveforms for the CT rectangular pulse and DT rectangular pulse are shown respectively, in figure 1.14.

Figure 1.14: Rectangular pulse

1.3.4 Signum function

The *signum* (or *sign*) function, denoted by $sgn(t)$, is defined as follows:

$$sgn(t) = \begin{cases} 1 & t > 0 \\ 0 & t = 0 \\ -1 & t < 0 \end{cases} \quad (1.10)$$

The DT signum function, denoted by $sgn(k)$, is defined as follows:

$$gn[k] = \begin{cases} 1 & k > 0 \\ 0 & k = 0 \\ -1 & k < 0 \end{cases} \quad (1.11)$$

The waveforms for the CT rectangular pulse and DT rectangular pulse are shown respectively, in figure 1.15.

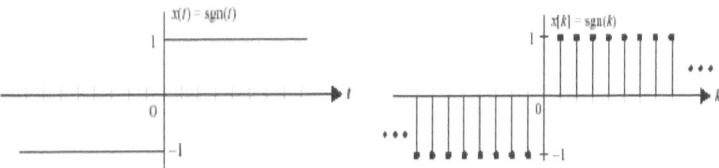

Figure 1.15: Signum function

1.3.5 Ramp function

The CT ramp function $r(t)$ is defined as follows:

$$r(t) = tu(t) = \begin{cases} t & t \geq 0 \\ 0 & t < 0 \end{cases} \quad (1.12)$$

Similarly, the DT ramp function r[k] is defined as follows:

$$r[k] = \begin{cases} k & k \geq 0 \\ 0 & k < 0 \end{cases} \quad (1.13)$$

The waveforms for the CT ramp function and DT ramp function are shown respectively, in Figure 1.16.

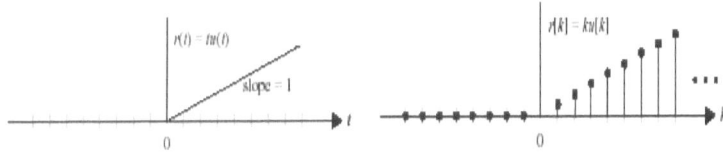

Figure 1.16 Ramp function

1.3.6 Sinc function

The CT *Sinc* function is defined as follows:

$$\text{sinc}(\omega_0 f) = \frac{\sin(\omega_0 f)}{\omega_0 f} \tag{1.14}$$

Similarly, the DT *Sinc* function is defined as follows:

$$\text{sinc}(\omega_0 k) = \frac{\sin(\omega_0 k)}{\omega_0 k} \tag{1.15}$$

The waveforms for the CT ramp function and DT ramp function are shown respectively, in figure 1.17

Figure 1.17 Sinc function

1.3.7 Exponential Function

Exponential signals are extremely important in signals and systems analysis because they are eigenfunctions of linear time-invariant systems (more on this later…)
- Real Exponential Signals Continuous-Time

$$x(t) = Ce^{at}, \quad C, a \text{ real} \tag{1.16}$$

Case $a = 0$: We simply get the constant signal $x(t) = C$.
Case $a > 0$: The exponential tends to infinity as $t \to \infty$ (here C>0).

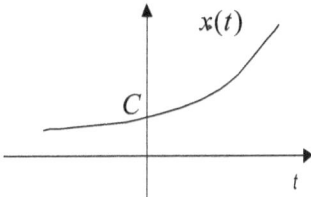

Case $a < 0$: The exponential tends to zero as $t \to \infty$ (here $C > 0$).

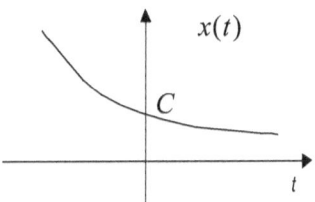

- Real Exponential Signals Discrete-Time

$$x[n] = C\alpha^n, \quad C, \alpha \text{ real} \tag{1.17}$$

There are six cases to consider (apart from the trivial case $a = 0$): $a = 1$, $a > 1$, $0 < a < 1$, $a < -1$, $a = -1$ and $-1 < a < 0$. Here we assume that $C > 0$.

Case $a = 1$: We get a constant signal $x[n] = C$.

Case $a > 1$: We get a positive signal that grows exponentially.

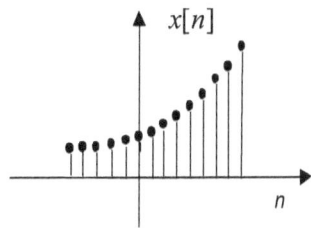

Case $0 < a < 1$: The signal is positive and decays exponentially.

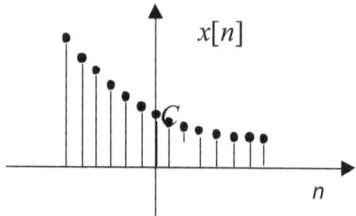

Case $a < -1$: The signal alternates between positive and negative values and grows exponentially.

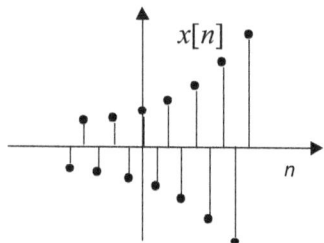

Case $a = -1$: The signal alternates between + C and - C.

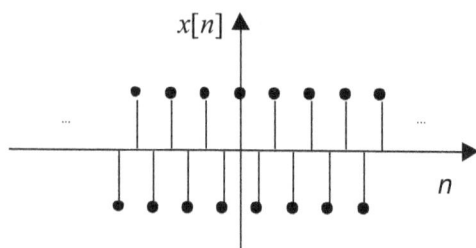

Case $-1 < a < 0$: The signal alternates between positive and negative values and decays exponentially.

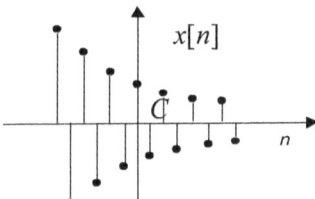

Problems

1.1 For each of the following representations:

   i. $z[m, n, k]$,
   ii. $I(x, y, z, t)$,

   Establish if the signal is a CT or a DT signal. Specify the independent and dependent variables. Think of an information signal from a physical process that follows the mathematical representation given in (i). Repeat for the representation in (ii).

1.2 Sketch each of the following CT signals as a function of the independent variable t over the specified range:

   i. $x1(t) = \cos(3t/4 + \pi/8)$ for $-1 \leq t \leq 2$;
   ii. $x2(t) = \sin(-3t/8 + \pi/2)$ for $-1 \leq t \leq 2$;
   iii. $x3(t) = 5t + 3\exp(-t)$ for $-2 \leq t \leq 2$;
   iv. $x4(t) = (\sin(3t/4 + \pi/8))^2$ for $-1 \leq t \leq 2$;
   v. $x5(t) = \cos(3t/4) + \sin(t/2)$ for $-2 \leq t \leq 3$;
   vi. $x6(t) = t \exp(-2t)$ for $-2 \leq t \leq 3$.

1.3 Sketch the following DT signals as a function of the independent variable k over the specified range:

   i. $x1[k] = \cos(3k/4 + \pi/8)$ for $-5 \leq k \leq 5$;
   ii. $x2[k] = \sin(-3k/8 + \pi/2)$ for $-10 \leq k \leq 10$;
   iii. $x3[k] = 5k + 3^{-k}$ for $-5 \leq k \leq 5$;
   iv. $x4[k] = |\sin(3k/4 + \pi/8)|$ for $-6 \leq k \leq 10$;
   v. $x5[k] = \cos(3k/4) + \sin(k/2)$ for $-10 \leq k \leq 10$;
   vi. $x6[k] = k4^{-|k|}$ for $-10 \leq k \leq 10$.

1.4 Determine if the following CT signals are periodic. If yes, calculate the fundamental period $T0$ for the CT signals:

i. $x1(t) = \sin(-5t/8 + /2)$;
ii. $x2(t) = |\sin(-5t/8 + /2)|$;
iii. $x3(t) = \sin(6t/7) + 2\cos(3t/5)$;
iv. $x4(t) = \exp(j(5t + /4))$;
v. $x5(t) = \exp(j3t/8) + \exp(t/86)$;
vi. $x6(t) = 2\cos(4t/5)^*\sin^2(16t/3)$;
vii. $x7(t) = 1 + \sin 20t + \cos(30t + /3)$.

1.5 Determine if the following DT signals are periodic. If yes, calculate the fundamental period $N0$ for the DT signals:

i. $x1[k] = 5 \times (-1)^k$
ii. $x2[k] = \exp(j(7k/4)) + \exp(j(3k/4))$;
iii. $x3[k] = \exp(j(7k/4)) + \exp(j(3k/4))$;
iv. $x4[k] = \sin(3k/8) + \cos(63k/64)$;
v. $x5[k] = \exp(j(7k/4)) + \cos(4k/7 +)$;
vi. $x6[k] = \sin(3k/8) \cos(63k/64)$.

1.6 Show that the average power of the CT periodic signal $x(t) = A \sin(\omega_0 t + \theta)$, with real-valued coefficient A, is given by $A^2/2$.

1.7 Show that the average power of the CT periodic signal $x(t) = D \exp[j(\omega_0 t + \theta)]$ is given by D^2.

1.8 Determine if the following CT signals are even, odd, or neither even nor odd. In the latter case, evaluate and sketch the even and odd components of the CT signals:

i. $x1(t) = 2\sin(2\_t)[2 + \cos(4\_t)]$;
ii. $x2(t) = t2 + \cos(3t)$;
iii. $x3(t) = \exp(-3t)\sin(3\_t)$;
iv. $x4(t) = t\sin(5t)$;

1.9 Determine if the following DT signals are even, odd, or neither even nor odd. In the latter case, evaluate and sketch the even and odd components of the DT signals:

i. $x1[k] = \sin(4k) + \cos(2\pi k/3)$;
ii. $x2[k] = \sin(\pi k/3000) + \cos(2\pi k/3)$;
iii. $x3[k] = \exp(j(7\pi k/4)) + \cos(4\pi k/7 + \pi)$;
iv. $x4[k] = \sin(3\pi k/8)\cos(63\pi k/64)$;

Chapter Two

Introduction to Systems

Learning Outcomes of this Chapter

After successful completion of this chapter students will be able to:

1. characterize and analyze the properties of Discrete systems.
2. understand system properties - linearity, time invariance, presence or absence of memory, causality, bounded-input bounded-output stability, and invertibility
3. perform the process of convolution between signals and understand its implication for analysis of linear time-invariant systems.

2.1 Introduction

The concept of a system is very similar to that of a signal. One of the most important distinctions to understand is the difference between discrete time and continuous time systems. A system in which the input signal and output signal both have continuous domains is said to be a continuous system. One in which the input signal and output signal both have discrete domains is said to be a discrete system. Of course, it is possible to conceive of signals that belong to neither category such as systems in which sampling of a continuous time signal or reconstruction from a discrete time signal take place A system is also a mapping between two sets; however, both the domain and the range of a system are sets of signals. A system is thus a rule for producing a signal in its range, given a signal from its domain.

Recall also that we classified signals according to their domains and ranges. For example, a signal whose domain is an interval of integers and whose range is an interval of reals is called a discrete-time signal. We can similarly classify systems. Specifically, we will distinguish two important classes of systems. For a discrete-time (DT) system, both the range and the domain are sets of DT signals. For a continuous-time (CT) system, both the range and the domain are sets of CT signals.

A system can be represented as a block diagram, as in Figure 2.1(a). It is important to remember that the input and output are not single numbers, they are signals. The specification of a range and a domain are crucial for defining both signals and systems. In fact, the actual mapping may be identical for a signal f and a system S; the two will however always have different ranges and domains: the range and the domain for f are sets of numbers while the range and the domain for S are sets of signals.

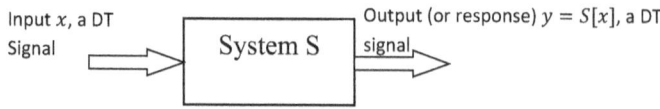

Figure 2.1 (a) A generic block diagram for a system

To illustrate this point, let us consider the example of the discrete-time function "divide by 3", shown in Figure 2.1 (b). The two objects are completely different: the function "divide by 3" takes in a single integer number n and produces a single real number n/3, whereas the system "divide by 3" takes in a DT signal x and produces another DT signal y such that y(n) = x(n)/3 for all integer values of n.

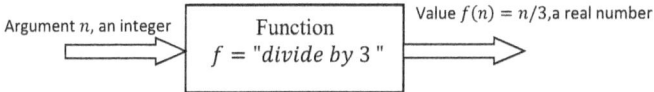

Figure 2.1 (b) DT signal "divide by 3"

A specific example of this is given in Figure 2.2: supposing that the input signal is $x(n) = cos(n)$, the output is another signal, $y(n) = cos(n)/3$. In other words, x is a rule for transforming a single number into another number; the system changes this rule into y.

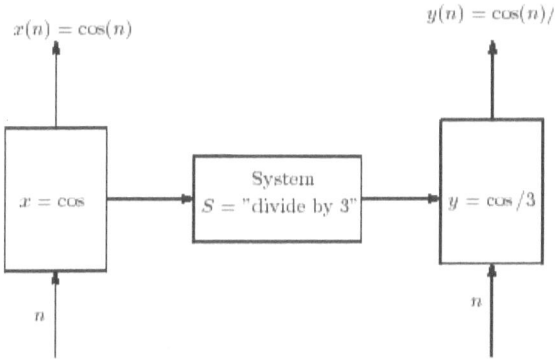

Figure 2.2 System "divide by 3" for the specific case when the input signal is $x = cos$

Another way of thinking about what a system does is that the whole graph of the input signal x is fed into S, and it produces the whole graph of the output signal y, as depicted in Figure 2.3. To emphasize that the input of S is the whole signal x, we will be using $S[x]$ to denote the output signal y, rather then $S[x(n)]$. The latter notation is also acceptable, provided you keep in mind that what it really stands for is: $S[x(n)$ for all $n]$, i.e. that S operates on all the samples of x Once the system's response is known, it can be evaluated at a particular n: $S[x](n)$ is synonymous with $y(n)$ and means the n-th sample of y, where y is the response of system S to input x.

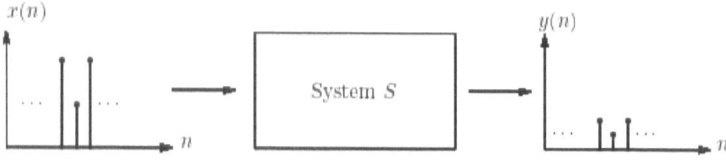

Figure 2.3 Another view of what a system is.

2.2 Classification of Systems

Every system can be characterized by its ability to accept an input such as voltage, pressure, etc. and to produce an output in response to this input. An example is a filter whose input is a signal corrupted by noise and interference and whose output is the desired signal. So, a system can be viewed as a process that results in transforming input signals into output signals.

First, we review the concept of systems by discussing the classification of systems according to the way the system interacts with the input signal. This interaction, which defines the model for the system, can be linear or nonlinear, time-invariant or time varying, memoryless or with memory, causal or noncausal, stable or unstable, and deterministic or nondeterministic. We briefly review the properties of each of these classes.

There are some fundamental properties that many (but not all!) systems share regardless if they are C-T or D-T and regardless if they are electrical, mechanical, etc.

An understanding of these fundamental properties allows an engineer to develop tools that can be widely applied... rather than attacking each seemingly different problem a new. There are six basic categories for systems:

i. Linear and non-linear systems.
ii. Time-invariant and time-varying systems.
iii. Systems with and without memory.
iv. Invertible and non-invertible systems.
v. Causal and non-causal systems.
vi. Stable and unstable systems.

2.2.1 Linear and non-linear systems

The class of linear systems is defined by the principle of superposition. If $y_1[n]$ and $y_2[n]$ are the response of a system when $x_1[n]$ and x2[n] are the respective inputs, then the system is linear if and only if

$$T\{x_1[n] + x_2[n]\} = T\{x_1[n]\} + T\{x_2[n]\} = y_1[n] + y_2[n] \quad (2.1)$$
and
$$T\{ax[n]\} = aT\{x[n]\} = ay[n], \quad (2.2)$$

where a is an arbitrary constant. The first property is the additivity property, and the second the homogeneity or scaling property. These two properties together comprise the principle of superposition, stated as:

$$T\{ax_1[n] + bx_2[n]\} = aT\{x_1[n]\} + bT\{x_2[n]\} \quad (2.3)$$

for arbitrary constants a and b. This equation can be generalized to the superposition of many inputs. Specifically, if

$$x[n] = k\, a_k x_k[n], \quad (2.4)$$

then the output of a linear system will be

$$y[n] = k\, a_k y_k[n], \quad (2.5)$$

where $y_k[n]$ is the system response to the input $x_k[n]$.

Example 2.1 Accumulator System

Accumulator System defined by the input–output equation

$$y[n] = \sum_{k=-\infty}^{n} x[k] \tag{2.6}$$

is called the accumulator system, since the output at time n is the accumulation or sum of the present and all previous input samples. The accumulator system is a linear system. Since this may not be intuitively obvious, it is a useful exercise to go through the steps of more formally showing this. We begin by defining two arbitrary inputs $x_1[n]$ and $x_2[n]$ and their corresponding outputs.

$$y_1[n] = \sum_{k=-\infty}^{n} x_1[k]$$

$$y_2[n] = \sum_{k=-\infty}^{n} x_2[k]$$

When the input is $x_3[n] = ax_1[n] + bx_2[n]$, the superposition principle requires the output $y_3[n] = ay_1[n] + by_2[n]$ for all possible choices of a and b. We can show this by starting from Eq. (2.6):

$$y_3[n] = \sum_{k=-\infty}^{n} x_3[k]$$

$$= \sum_{k=-\infty}^{n} (ax_1[k] + bx_2[k])$$

$$= a \sum_{k=-\infty}^{n} x_1[k] + b \sum_{k=-\infty}^{n} x_2[k]$$
$$= ay_1[n] + by_2[n]. \tag{2.7}$$

Thus, the accumulator system of Eq. (2.7) satisfies the superposition principle for all inputs and is therefore linear.

Example 2.2 A Nonlinear System
Consider the system defined by
$$w[n] = log10\,(|x[n]|)$$

The system is not linear. To prove this, we only need to find one counterexample that is, one set of inputs and outputs which demonstrates that the system violates the superposition principle. The inputs $x_1[n] = 1$ and $x_2[n] = 10$ are a counterexample. However, the output for $x_1[n] + x_2[n] = 11$ is:

$$log10(1 + 10) = log10(11) = log10(1) + log10(10) = 1.$$

Also, the output for the first signal is $w_1[n] = 0$, whereas for the second, $w_2[n] = 1$. The scaling property of linear systems requires that, since $x_2[n] = 10x_1[n]$, if the system is linear, it must be true that $w_2[n] = 10w_1[n]$. Since this is not so for Eq. (2.7) for this set of inputs and outputs, the system is not linear.

2.2.2 Time-varying and time-invariant systems

A time-invariant system (often referred to equivalently as a shift-invariant system) is a system for which a time shift or delay of the input sequence causes a corresponding shift in the output sequence. Specifically, suppose that a system transforms the input sequence with values $x[n]$ into the output sequence with values $y[n]$. Then, the system is said to be time invariant if, for all n_0, the input sequence with values $x_1[n] = x[n - n_0]$ produces the output sequence with values $y_1[n] = y[n - n0]$.

Therefore, a DT system with $x[k] \to y[k]$ is time-invariant if
$$x[k - n_0] \to y[k - n_0]$$
for any arbitrary discrete shift n_0.

Example 2.3
Accumulator as a Time-Invariant System

Consider the accumulator from Example 2.2. We define $x_1[n] = x[n - n_0]$. To show time invariance, we solve for both $y[n - n_0]$ and $y_1[n]$ and compare them to see whether they are equal. First,

$$y[n - n_0] = \sum_{k=-\infty}^{n-n_0} x[k] \qquad (2.8)$$

Next, we find

$$y_1[n] = \sum_{k=-\infty}^{n} x_1[k] \qquad (2.9)$$

$$= \sum_{k=-\infty}^{n} x[k - n_0] \qquad (2.10)$$

Substituting the change of variables $k_1 = k - n_0$ into the summation gives

$$y_1[n] = \sum_{k_1=-\infty}^{n-n_0} x[k_1] \qquad (2.11)$$

Since the index k in Eq. (2.8) and the index k_1 in Eq. (2.11) are dummy indices of summation, and can have any label, Eqs. (2.8) and (2.11) are equal and therefore $y_1[n] = y[n - n0]$. The accumulator is a time-invariant system.

The following example illustrates a system that is not time invariant.

Example 2.4

A Compressor System defined by the relation

$$y[n] = x[Mn], -\infty < n < \infty, \quad (2.12)$$

with M a positive integer, is called a compressor. Specifically, it discards $(M-1)$ samples out of M; i.e., it creates the output sequence by selecting every M^{th} sample. This system is not time invariant. We can show that it is not by considering the response $y_1[n]$ to the input $x_1[n] = x[n - n_0]$. For the system to be time invariant, the output of the system when the input is x1[n] must be equal to $y[n - n_0]$. The output $y_1[n]$ that results from the input $x_1[n]$ can be directly computed from Eq. (2.12) to be:

$$y_1[n] = x_1[n][Mn] = x[Mn - n_0]. \quad (2.13)$$

Delaying the output y[n] by n_0 samples yields

$$y[n - n0] = x[M(n - n_0)]. \quad (2.14)$$

Comparing these two outputs, we see that $y[n - n_0]$ is not equal to $y_1[n]$ for all M and $n0$, and therefore, the system is not time invariant.

It is also possible to prove that a system is not time invariant by finding a single counterexample that violates the time-invariance property. For instance, a counterexample for the compressor is the case when $M = 2$, $x[n] = \delta[n]$, and $x_1[n] = \delta[n - 1]$. For this choice of inputs and M, $y[n] = \delta[n]$, but $y_1[n] = 0$; thus, it is clear that $y_1[n] = y[n - 1]$ for this system.

2.2.3 Static and Dynamic Systems

A DT system is said to be static (memoryless) if its output $y[n]$ at instant n = n0 depends only on the value of its input $x[n]$ at the same instant $n = n_0$. Otherwise, the DT system is Dynamic (said to have memory). For example, the system specifies by the input-output relationship:

$$y[n] = (2x[n] - x^2[n]) \tag{2.15}$$

is static system as the values of $y[n]$ at any particular time n0 depends only on the values of $x[n]$ at that time. Similarly, a resistor is a memoryless system: with the input x(t) taken as the current and with the voltage taken as an output y(t), the input-output relationship of a resistor is:

$$y(t) = Rx(t) \tag{2.16}$$

where R is the resistor. One particularly simple memoryless system is the identity system, whose output is identical to its input. That is, the input-output relationship for the continuous-time identity system is

$$y(t) = x(t),$$

and the corresponding relationship in discrete time is

$$y[n] = x[n].$$

An example of a discrete-time system with memory is an accumulator or summer.

$$y[n] = \sum_{k=-\infty}^{n} x[k] \tag{2.17}$$

and a second example is a delay

$$y[n] = x[n-1] \tag{2.18}$$

A capacitor is an example of a continuous-time system with memory, since if the input is taken to be the current and the output is the voltage, then

$$y[t] = \frac{1}{c}\int_{-\infty}^{T} x(\tau)d\tau \qquad (2.19)$$

where C is the capacitance. Roughly speaking, the concept of memory in a system corresponds to the presence of a mechanism in the system that retains or stores information about input values at times other than the current time.

2.2.4 Invertible and non-invertible systems

A system is said to be invertible if distinct inputs lead to distinct outputs. Said another way, a system is invertible if by observing its output; we can determine its input. For the discrete–time case, we can construct an inverse system which when cascaded with the original system yields an output z[n] equal to the input x[n] to the first system. Thus, the series interconnection in Figure 2.4 has an overall input – output relationship that is the same as that for the identity system.

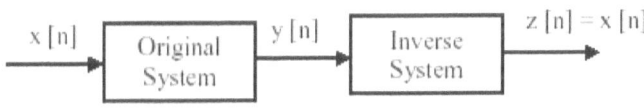

Figure 2.4

For example, the difference between two successive values of the output is precisely the last input value. Therefore, in this case, the inverse system is:

$$w[n] = y[n] - y[n-1], \qquad (2.20)$$

The concept of invertibility is important in many contexts. One example arises in systems for encoding used in a wide variety of communications applications. In such a system, a signal that we wish to transmit is first applied as the input to a system known as an encoder. There are many reasons for doing this, ranging from the desire to encrypt the original message for secure or private communication to the objective of providing some redundancy in the signal (for example, by adding what are known as parity bits) so that any errors that occur in transmission can be detected and, possibly, corrected. For lossless coding, the input to the encoder must be exactly recoverable from the output; i.e., the encoder must be invertible.

2.2.5 Causal and non-causal systems

A system is causal if the output at any time depends only on values of the input sequence at the present and past time. This implies that if $x_1[n] = x_2[n]$ for $n \leq n_0$, then $y_1[n] = y_2[n]$ for $n \leq n0$. That is, the system is nonanticipative.

Example 2.5
The Forward and Backward Difference Systems The system defined by the relationship

$$y[n] = x[n + 1] - x[n] \qquad (2.21)$$

is referred to as the forward difference system. This system is not causal, since the current value of the output depends on a future value of the input. The violation of causality can be demonstrated by considering the two inputs $x_1[n] = \delta[n - 1]$ and $x_2[n] = 0$ and their corresponding outputs $y_1[n] = \delta[n] - \delta[n - 1]$ and $y_2[n] = 0$ for all n. Note that $x_1[n] = x_2[n]$ for $n \leq 0$, so the definition of causality requires that $y_1[n] = y_2[n]$

for n ≤ 0, which is clearly not the case for $n = 0$. Thus, by this counterexample, we have shown that the system is not causal.

The backward difference system, defined as

$$y[n] = x[n] - x[n-1], \qquad (2.22)$$

has an output that depends only on the present and past values of the input. Because $y[n_0]$ depends only on $x[n_0]$ and $x[n_0 - 1]$, the system is causal by definition.

Example 2.6
When checking the causality of a system, it is important to look carefully at the input-output relation. To illustrate some of the issues involved in doing this, we will check the causality of two particular systems.

The first system is defined by

$$y[n] = x[-n]. \qquad (2.23)$$

Note that the output $y[n_0]$ at a positive time no depends only on the value of the input signal x[- no] at time ($-n_0$), which is negative and therefore in the past of no. We may be tempted to conclude at this point that the given system is causal. However, we should always be careful to check the input-output relation for all times. In particular, for $n < 0$, e.g. $n = -4$, we see that $y[-4] = x[4]$, so that the output at this time depends on a future value of the input. Hence, the system is not causal.

It is also important to distinguish carefully the effects of the input from those of any other functions used in the definition of the system. For example, consider the system

$$y(t) = x(t) \cos(t+1) \qquad (2.24)$$

In this system, the output at any time t equals the input at that same time multiplied by a number that varies with time. Specifically, we can rewrite eq. (2.24) as

$$y(t) = x(t)g(t),$$

where $g(t)$ is a time-varying function, namely $g(t) = cos(t + 1)$. Thus, only the current value of the input x(t) influences the current value of the output $y(t)$, and we conclude that this system is causal (and, in fact, memoryless).

2.2.6 Stable and unstable systems

Stability is another important system property. Informally, a stable system is one in which small inputs lead to responses that do not diverge.

Throughout this text, we specifically use bounded-input bounded-output stability. A system is stable in the bounded-input, bounded-output (BIBO) sense if and only if every bounded input sequence produces a bounded output sequence. The input x[n] is bounded if there exists a fixed positive finite value Bx such that

$$|x[n]| \leq Bx < \infty, \text{ for all n.}$$

Stability requires that, for every bounded input, there exists a fixed positive finite value By such that

$$|y[n]| \leq By < \infty, \text{ for all n.}$$

It is important to emphasize that the properties we have defined in this section are properties of systems, not of the inputs to a system. That is, we may be able to find inputs for which the properties hold, but the existence of the property for some inputs does not mean that the system has the property. For the system to have the property, it must hold for all inputs. For example, an unstable system may have some bounded inputs for which the output is bounded, but for the system to have the property of stability, it must be true that for all bounded inputs, the output is bounded. If we can find just one input for which the system property does not hold, then we have shown that the system does not have that property. The following example illustrates the testing of stability for several of the systems that we have defined.

Example 2.7 Testing for Stability or Instability
Testing for Stability or Instability The system of **square system** is stable. To see this, assume that the input $x[n]$ is bounded such that $|x[n]| \leq B_x$ for all n. Then $|y[n]| = |x[n]|^2 \leq B_x^2$. Thus, we can choose $By = B_x^2$ and prove that $y[n]$ is bounded.

Likewise, we can see that the system defined in Example 2.2 is unstable, since $y[n] = log10(|x[n]|) = -\infty$ for any values of the time index n at which $x[n] = 0$, even though the output will be bounded for any input samples that are not equal to zero.

The accumulator, as defined in Example 2.3 is also not stable. For example, consider the case when $x[n] = u[n]$, which is clearly bounded by $B_x = 1$. For this input, the output of the accumulator is

$$y[n] = \sum_{k=-\infty}^{n} u[k]$$
$$= \begin{cases} 0, & n < 0 \\ (n+1), & n \geq 0 \end{cases}$$

There is no finite choice for By such that $(n + 1) \leq By < \infty$ for all n; thus, the system is unstable.

2.3 Impulse Response and Convolution

We now take a closer look at LTI systems in the input-output form and develop a method to compute the output of an LTI system, given its input. Specifically, we will see that the output is the convolution of the input with the impulse response. Our plan for deriving this fact is:

1. Write the input signal as a linear combination (weighted sum) of shifted unit impulse signals.
2. Use linearity to write the response as the sum of responses to shifted impulses.

Use time-invariance to find the response to a shifted impulse. Specifically, the response of a time-invariant system to signal $\delta(n - k)$ is $h(n - k)$ where $h(n)$ is the unit impulse response.

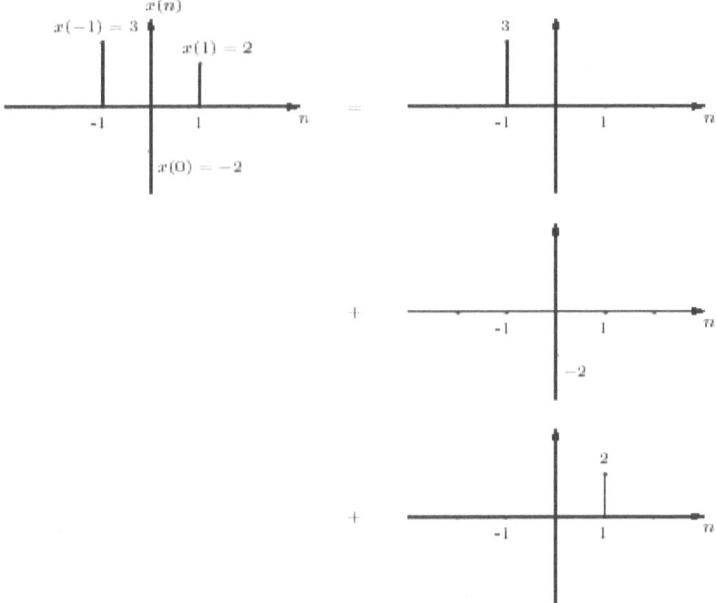

Figure 2.5. The signal x(n) is represented as a sum of impulse signals.

Let us begin with signal $x(n) = 3\delta(n+1) - 2\delta(n) + 2\delta(n-1)$, defined for all integer n. As shown in Figure 1.15, this signal can be represented as follows:

$$x(n) = x(-1)\delta_{-1}(n) + x(0)\delta_0(n) + x(1)\delta_1(n),$$

where δ_k is the unit impulse shifted by k, i.e. $\delta_k(n) = \delta(n-k)$ for all integer n and k. Similarly, any arbitrary signal can be represented as the following weighted sum of shifted impulse signals:

$$x(n) = \ldots + x(-2)\delta_{-2}(n) + x(-1)\delta_{-1}(n) + x(0)\delta_0(n) + x(1)\delta_1(n) + x(2)\delta_2(n) + \ldots$$
$$= \sum_{k=-\infty}^{\infty} x(k)\delta_k(n), \text{ for all integer } n.$$

If signal x is put through a linear system S, we can use the above equation and the linearity of the system to write the response y of the system as follows:

$y(n) = S[x](n)$
$= S[\ldots + x(-2)\delta_{-2} + x(-1)\delta_{-1} + x(0)\delta_0 + x(1)\delta_1 + x(2)\delta_2 + \ldots](n)$
$linearity = \ldots + x(-2)S[\delta_{-2}](n) + x(-1)S[\delta_{-1}](n) + x(0)S[\delta_0](n)$
$+ x(1)S[\delta_1](n) + x(2)S[\delta_2](n) + \ldots$

$$= \sum_{k=-\infty}^{\infty} x(k) S[\delta_k](n)$$

$$= \sum_{k=-\infty}^{\infty} x(k) h_k(n) \qquad (2.25)$$

where we denoted by $h_k = S[\delta_k]$ the system's response to the shifted impulse δ_k.

If system S, in addition to being linear, is time-invariant, then;

$$h_k(n) = h(n-k) \text{ for all integer } n \text{ and } k,$$

where h is the response to the unit impulse δ. Substituting this into Eq. (2.25) yields:

$$y(n) = \sum_{k=-\infty}^{\infty} x(k) h(n-k)$$

which is the formula for the discrete-time convolution. We will use the following notation to indicate that signal y is the convolution of signal x with signal h: $y = x * h$. The n-th sample of y is then $y(n) = x * h(n)$.

We have thus shown that the output of a discrete-time LTI system is the discrete time convolution of the input and the impulse response.

Example 2.8

Consider the following input-output specification of a system:

$y(n) = x(n) + \frac{1}{2}x(n-1)$, for all integer n:

Let us find the response to $x(n) = \begin{cases} \frac{1}{2}, & n = -1 \\ 1, & n = 0 \\ \frac{2}{3}, & n = 0 \\ 0, & \text{otherwise} \end{cases}$

The impulse response h of the system is the response to the unit impulse:

$$h(n) = \delta(n) + \frac{1}{2}\delta(n-1) = \begin{cases} 1, & n = 0, \\ \frac{1}{2}, & n = 1, \\ 0, & \text{otherwise} \end{cases}$$

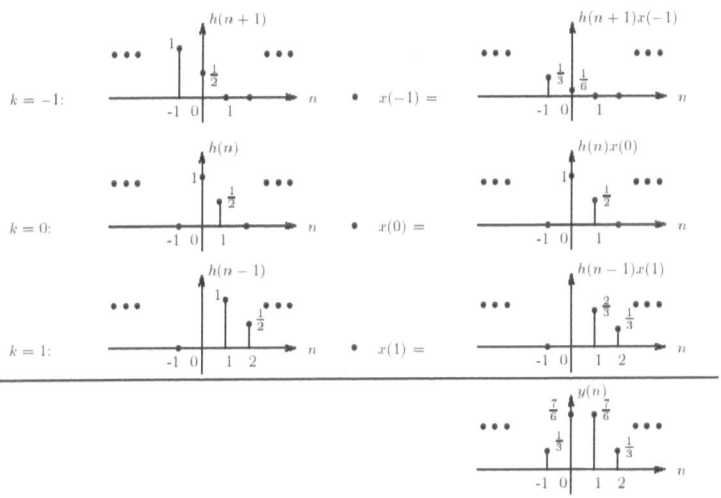

Figure 2.6. Illustration to Example 2.7: the convolution of h and x is a weighted linear combination of shifted versions of h, with the weights given by the samples of x.

Therefore, the response to x is:

$$y(n) = \sum_{k=-\infty}^{\infty} x(k)h(n-k) = x(-1)h(n+1) + x(0)h(n) + x(1)h(n-1),$$

for all integer n.

We can evaluate this convolution by directly calculating the linear combination of shifted versions of h. We start by plotting $h(n-k)$ as a function of n, for each k, and proceed as shown in Figure 2.6.

A more convenient method is illustrated in Figure 2.7. It involves plotting signal $x(k)$ as a function of k, and plotting signals $h(n-k)$ as a function of k, for each n. Here is the basic procedure for calculating the $n-th$ sample of y:

(1) flip h;
(2) for a fixed n, shift h by n;
(3) for the same fixed n, multiply $x(k)$ by $h(n-k)$, for each k;
(4) Sum the products over k: $\sum_{k=-\infty}^{\infty} x(k)h(n-k) = y(n)$

APPLIED DIGITAL SIGNAL PROCESSING AND APPLICATIONS

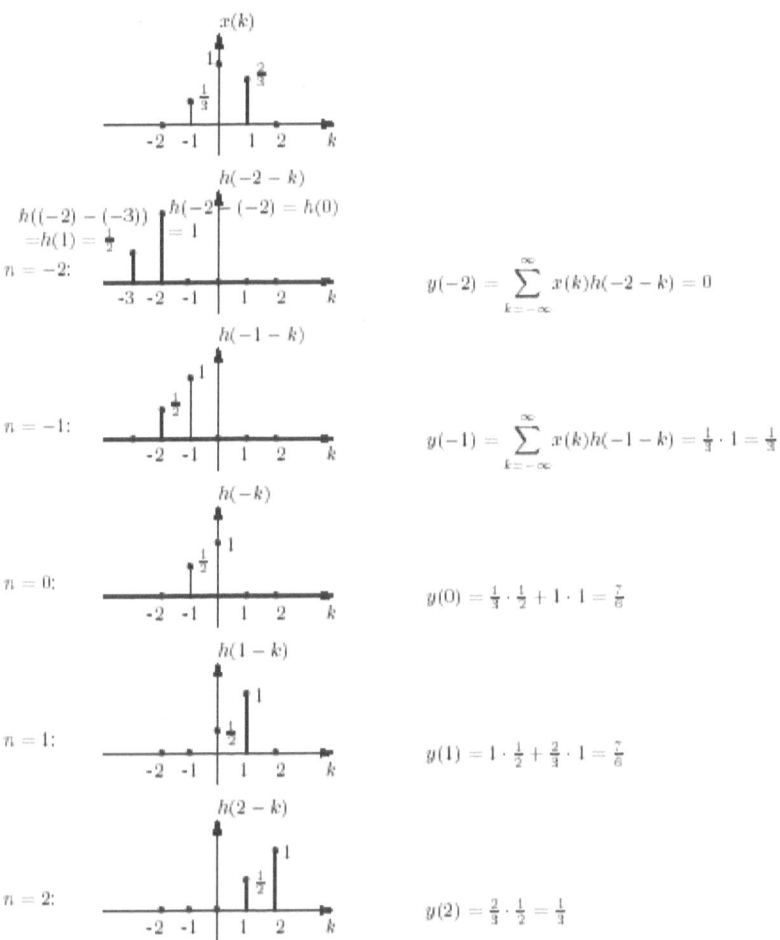

Figure 2.7. Evaluating the convolution sum of Example 1.6.

Both methods of course lead to the same result:

$$y(n) = \begin{cases} \dfrac{1}{2}, & n = -1, 2 \\ \dfrac{7}{6}, & n = 0, 1 \\ 0, & \text{otherwise} \end{cases}$$

Example 2.9

To evaluate the convolution of signals $x(n) = 2^{-|n|}$ and $h(n) = u(n)$, we substitute the two expressions into the definition of convolution:

$$y(n) = x * h(n) = \sum_{k=-\infty}^{\infty} 2^{-|n|} u(n-k)$$

$$= \sum_{k=-\infty}^{n} 2^{-|k|}$$

For $n \leq 0$, the summation is only over nonpositive values of k and therefore $|k|$ can be replaced with $-k$:

$$y(n) = \sum_{k=-\infty}^{n} 2^{-(-k)} = \sum_{k=-\infty}^{n} 2^{k}$$

$$= \sum_{m=-n}^{\infty} 2^{-m} = \sum_{m=-n}^{\infty} (\frac{1}{2})^m = \frac{(\frac{1}{2})^{-n}}{1 - 1/2}$$

$$= 2^{n+1}, \text{ for any integer } n \leq 0,$$

where we substituted $m = -k$. When $n > 0$, the summation can be broken into two pieces: one for nonpositive values of k (i.e. for k from $-\infty$ to 0) and the other for positive values of k (i.e. for k from 1 to n):

$$y(n) = \sum_{k=-\infty}^{n} 2^{-(-k)} = \sum_{k=-1}^{n} 2^{-k}$$

$$= \sum_{m=0}^{\infty} 2^{-m} = \sum_{k=1}^{n} 2^{-k}$$

$$= \frac{1}{1-1/2} + \frac{(\frac{1}{2})^1 - (\frac{1}{2})^{n+1}}{1-1/2}$$

$$= 3 - 2^{-n}, \text{ for any integer } n > 0.$$

Putting together the two cases,

$$y(n) = \begin{cases} 2^{n+1}, & n \leq 0 \\ 3 - 2^{-n}, & n > 0 \end{cases}$$

Problems

2.1. For each of the following systems, determine whether the system is (1) stable, (2) causal, (3) linear, (4) time invariant, and (5) memoryless:

i. $T(x[n]) = g[n]x[n]$ with $g[n]$ given.

ii. $T(x[n]) = \sum_{kn_0}^{n} x[k] \quad n \neq 0$

iii. $T(x[n]) = \sum_{kn-n_0}^{n+n_0} x[k]$

iv. $T(x[n]) = x[n - n_0]$

v. $T(x[n]) = e^{x[n]}$

vi. $T(x[n]) = ax[n] + b$

vii. $T(x[n]) = x[-n]$

viii. $T(x[n]) = x[n] + 3u[n + 1]$.

2.2. Determine the output of an LTI system if the impulse response $h[n]$ and the input $x[n]$ are as follows:

i. $x[n] = u[n]$ and $h[n] = a^n u[-n - 1]$, with $a > 1$.

ii. $x[n] = u[n - 4]$ and $h[n] = 2^n u[-n - 1]$.

iii. $x[n] = u[n]$ and $h[n] = (0.5)2^n u[-n]$.

iv. $h[n] = 2^n u[-n - 1]$ and $x[n] = u[n] - u[n - 10]$.

Use your knowledge of linearity and time invariance to minimize the work in parts (ii)–(iv).

2.3. Consider the input sequence $x[k] = 2u[k]$ applied to a DT system modelled with the following input–output relationship:

$$y[k + 1] - 2y[k] = x[k],$$

and ancillary condition $y[-1] = 2$.

i. Determine the response $y[k]$ by iterating the difference equation for $0 \leq k \leq 5$.
ii. Determine the zero-state response $y_{zi}[k]$ for $0 \leq k \leq 5$
iii. Calculate the zero-input response $y_{zs}[k]$ for $0 \leq k \leq 5$.
iv. Verify that $y[k] = y_{zi}[k] + y_{zs}[k]$.

2.4. Calculate the convolution $(x_1[k] * x_2[k])$ for the following pairs of sequences:

i. $x_1[k] = u[k+2] - u[k-3]$, $x_2[k] = u[k+4] - u[k-5]$;
ii. $x_1[k] = 0.5^k u[k]$, $x_2[k] = 0.8^k u[k-5]$;
iii. $x_1[k] = 7^k u[-k+2]$, $x_2[k] = 0.4^k u[k-4]$;
iv. $x_1[k] = 0.6^k u[k]$, $x_2[k] = \sin(\pi k/2) u[-k]$;
v. $x_1[k] = 0.5^{|k|}$, $x_2[k] = 0.8^{|k|}|$.

2.5. Consider a discrete-time system with input $x[n]$ and output $y[n]$. The input-output relationship for this system is

$$y[n] = x[n]x[n-2].$$

i. Is the system memoryless?
ii. Determine the output of the system when the input is $A\delta[n]$ where A is any real or complex number.
iii. Is the system invertible?

2.6. Consider a discrete-time system with input $x[n]$ and output $y[n]$ related by

$$y[n] = \sum_{k=n-n_0}^{n+n_0} x[k],$$

where n_0 is a finite positive integer.

i. Is this system linear?
ii. Is this system time-invariant?
iii. If $x[n]$ is known to be bounded by a finite integer B (i.e., $|x[n[|< B$ for all n), it can be shown that $y[n]$ is bounded by a finite number C. We conclude that the given system is stable. Express C in terms of B and n_0.

2.7 An LTI system is described by the input–output relation

$$y[n] = x[n] + 2x[n-1] + x[n-2].$$

i. Determine $h[n]$ the impulse response of the system.
ii. Is this a stable system?
iii. Determine $H(e^{jw})$, the frequency response of the system. Use trigonometric identities to obtain a simple expression for $H(e^{jw})$.
iv. Plot the magnitude and phase of the frequency response.
v. Now consider a new system whose frequency response is $H_1(e^{jw}) = H(e^{j(\omega+\pi)}H)$. Determine $n_1[n]$, the impulse response of the new system.

2.8 Consider a system with input x[n] and output y[n]. The input–output relation for the system is defined by the following two properties:

1. $y[n] - ay[n-1] = x[n]$,
2. $y[0] = 1$.

i. Determine whether the system is time invariant.
ii. Determine whether the system is linear.
iii. Assume that the difference equation (property 1) remains the same, but the value $y[0]$ is specified to be zero. Does this change your answer to either part **(i)** or part **(ii)**?

Chapter Three

Sampling, Quantization and Reconstruction

Learning Outcomes of this Chapter

After successful completion of this chapter students will be able to:

1. convert an analog signal to a discrete-time sequence via sampling
2. construct an analog signal from a discrete-time sequence.
3. understanding the conditions when a sampled signal can uniquely present its analog counterpart.
4. understand the Nyquist sampling theorem and the process of reconstructing a continuous-time signal from its samples.

3.1 Introduction

Often the domain and the range of an original signal $x(t)$ are modeled as continuous. That is, the time (or spatial) coordinate t is allowed to take on arbitrary real values (perhaps over some interval) and the value $x(t)$ of the signal itself is allowed to take on arbitrary real values (again perhaps within some interval). As mentioned previously in Chapter one, such signals are called analog signals. A continuous model is convenient for some situations, but in other situations it is more convenient to work with digital signals i.e., signals which have a discrete (often finite) domain and range. The process of digitizing the domain is called sampling and the process of digitizing the range is called quantization. Most devices we encounter deal with both analog and digital signals. Digital signals are particularly robust to noise, and extremely efficient and versatile means for processing digital signals have been developed. On the other hand, in certain situations analog signals are sometimes more appropriate or even necessary. For example, most underlying physical processes are analog (or at least most conveniently modeled as analog), including the human sensorimotor systems. Hence, analog signals are typically necessary to interface with sensors and actuators. Also, some types of data processing and transmission are most conveniently performed with analog signals. Thus, the conversion of analog signals to digital signals (and vice versa) is an important part of many information processing systems.

Before we sample any signal, we have to filter the signal to limit the maximum frequency of the signal as it affects the sampling rate. Filtering should ensure that we do not distort the signal, ie remove high frequency components that affect the signal shape. Sampling is simply the process of measuring the value of a continuous-time signal at certain instants of time. Typically, these measurements are uniformly separated by the sampling period, T_s. If $x(t)$ is the input signal, then the sampled signal, $y(n)$, is as follows: $y(n) = x(t)| t = nT_s$. A critical question is the following: What sampling period, T_s, is required to accurately represent the signal $x(t)$? To answer this question, we need to look at the frequency domain representations of $y(n)$ and $x(t)$.

3.2 Signal Sampling

The sampling results in a discrete set of digital numbers that represent measurements of the signal - usually taken at equal intervals of time. Note that the sampling takes place after the hold. This means that we can sometimes use a slower Analogue to Digital Converter (ADC) than might seem required at first sight. The hold circuit must act fast - fast enough that the signal is not changing during the time the circuit is acquiring the signal value - but the ADC has all the time that the signal is held to make its conversion. A gain, note that at this point (after sampling), our signal is not yet completely digital because the values $x[n]$ can still take on any number from a continuous range - that's why we use the terms discrete-time signal here and not digital signal. Figure 3.1 illustrates the process of sampling a continuous sinosoid. Although it has been drawn in the right plot, the underlying continuous signal is lost in this process - all we have left after the sampling is a sequence of numbers. Those numbers themselves are termed samples in the DSP community - each such number is a sample in this terminology. This is different from what a musician usually means when talking about samples - musicians refer to a short recording of an acoustic event as sample.

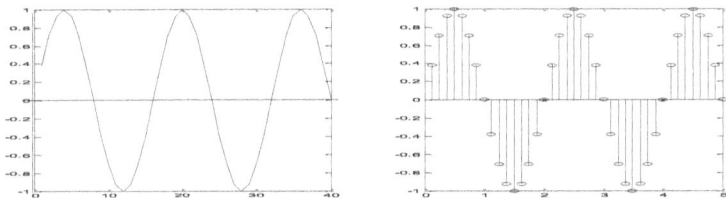

Figure 3.1 Sampling a sinosoid

The samples shown are equally spaced and simply pick off the value of the underlying analog signal at the appropriate times. If we let T denote the time interval between samples, then

the times at which we obtain samples are given by nT where $n = \ldots, -2, -1, 0, 1, 2, \ldots$. Thus, the discrete-time (sampled) signal $x[n]$ is related to the continuous-time signal by
$$x[n] = x(nT).$$

It is often convenient to talk about the sampling frequency f_s. If one sample is taken every T seconds, then the sampling frequency is $f_s = 1/T$ Hz. The sampling frequency could also be stated in terms of radians, denoted by ω_s. Clearly, $\omega_s = 2\pi f_s = 2\pi/T$.

The type of sampling mentioned above is sometimes referred to as "ideal" sampling. In practice, there are usually two non-ideal effects. One effect is that the sensor (or digitizer) obtaining the samples can't pick off a value at a single time. Instead, some averaging or integration over a small interval occurs, so that the sample actually represents the average value of the analog signal in some interval. This is often modeled as a convolution – namely, we get samples of $y(t) = x(t) * h(t)$, so that the sampled signal is $y[n] = y(nT)$. In this case, $h(t)$ represents the impulse response of the sensor or digitizer. Actually, sometimes this averaging can be desirable. For example, if the original signal $x(t)$ is changing particularly rapidly compared to the sampling frequency or is particularly noisy, then obtaining samples of some averaged signal can actually provide a more useful signal with less variability. The second non-ideal effect is noise. Whether averaged or not, the actual sample value obtained will rarely be the exact value of the underlying analog signal at some time. Noise in the samples is often modeled as adding (usually small) random values to the samples.

Although in real applications there are usually non-ideal effects such as those mentioned above, it is important to consider what can be done in the ideal case for several reasons.

The non-ideal effects are often sufficiently small that in many practical situations they can be ignored. Even if they cannot be ignored, the techniques for the ideal case provide insight into how one might deal with the non-ideal effects. For simplicity, we will usually assume that we get ideal, noise-free samples.

3.3 Interpolation

Whereas the continuous signal $x(t)$ is defined for all values of t, our discrete-time signal is only defined for times which are integer multiples of T_s. To reconstruct a continuous signal from the samples, we must somehow 'guess', what value the signal could probably take on in between our samples. Interpolation is the process of 'guessing' signal values at arbitrary instants of time, which fall - in general - in between the actual samples. Thereby interpolation creates a continuous time signal and can be seen as an inverse process to sampling. Ideally, we would want our interpolation algorithm to 'guess right' - that is: the continuous signal obtained from interpolation should be equal to the original continuous signal. The crudest of all interpolation schemes is piecewise constant interpolation - we just take the value of one of the neighbouring samples as guessed signal value at any instant of time in between. The reconstructed interpolated function will have a stairstep-like shape. The next better and very popular interpolation method is linear interpolation - to reconstruct a signal value we simply connect the values at our sampling instants with straight lines. Figure 3.2 shows the reconstructed continuous signals for piecewise constant.

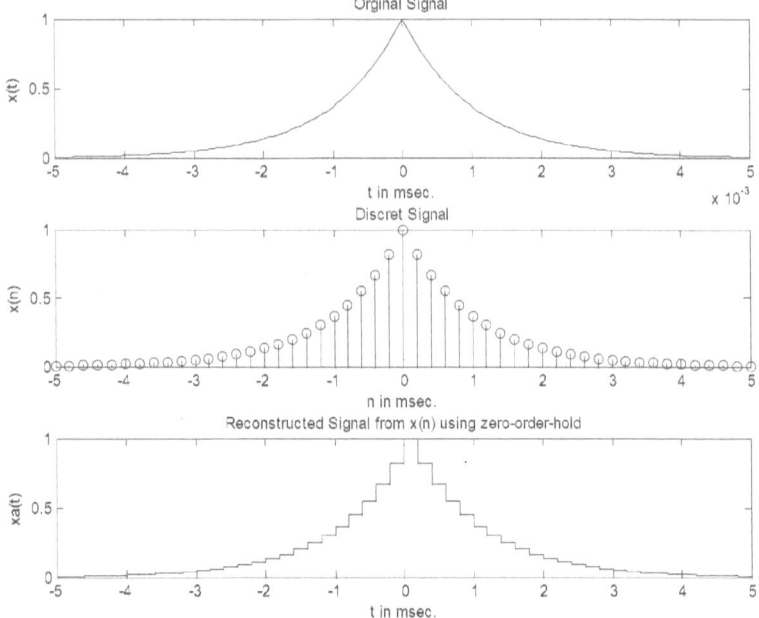

Figure 3.2 Signal reconstruction

3.4 The Sampling Theorem

An analog signal can be reconstructed from its sampled values un-erroneously, if the sampling frequency is at least twice the bandwidth of the analog signal. Suppose we sample a band-limited signal and choose the sampling frequency such that $f_{sam} \geq 2f_m$. The spectra of the continuous-time signal and of the sampled signal are exemplified in Figure 3.3. Notice that in this case the replicas in the sampled signal do not overlap. This is the principle of the Nyquist rate of sampling.

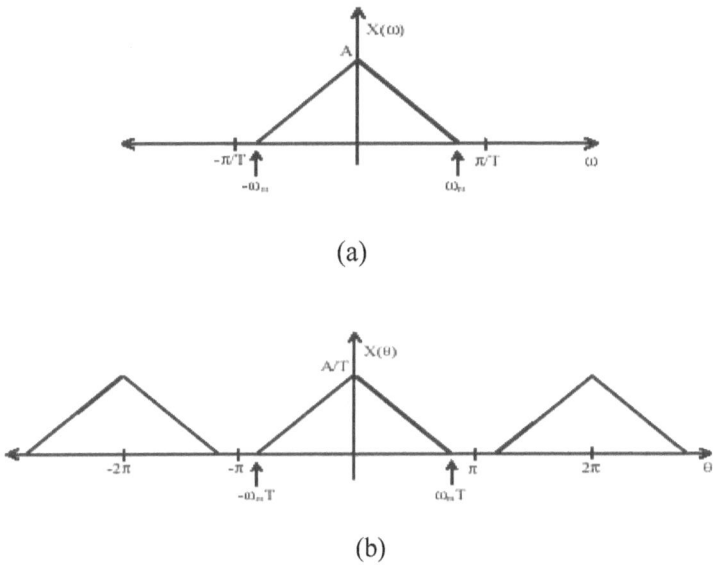

Figure 3.3 Sampling of a band limited signal above the Nyquist rate: (a) Fourier transform of the continuous-time signal; (b) Fourier transform of the sampled signal.

The *Nyquist sampling rate*, or the critical sampling rate for a band limited signal is mathematically defined by Equation 3.1, where f_m is the maximum frequency of the signal.

$$f_{sam} = \frac{1}{T} = 2f_m \qquad (3.1)$$

By definition, if a band limited signal is sampled at a rate equal to or greater than the Nyquist sampling frequency $2f_m$. Then the shape of the Fourier transform of the sampled signal in the range [-π, π] is identical to the shape of the Fourier transform of the given signal, except for multiplication of the frequency axis by a factor T, and multiplication of the amplitude axis by a factor $1/T$, as shown in Figure 3.3(b).

3.5 Aliasing

Aliasing is caused by sampling at a rate lower than that of the Nyquist frequency for a given signal. It is an effect that occurs when a signal is sampled at too low a frequency. What happens is that the higher frequency components of the signal cannot be captured because of the low sampling frequency, which results in overlap in the spectrum.

Now let us consider the case of a band limited signal when it is sampled at a lower rate than the Nyquist frequency. As an example, Figure 3.4 illustrates the case of sampling at $f_{sam} = 3f_m/2$. Here, the shape of the Fourier transform of the sampled signal in the range $[-\pi, \pi]$ becomes distorted. Distortion occurs in this frequency range because two adjacent replicas overlap and their superposition give rise to the shape illustrated by Figure 3.4 (b).

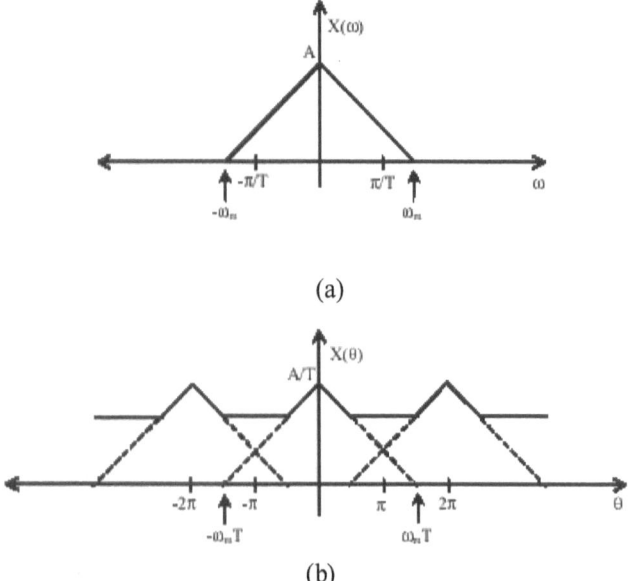

Figure 3.4. Sampling of a band limited signal below the Nyquist rate: (a) Fourier transform of the continuous-time signal; (b) Fourier transform of the sampled signal.

3.6 Antialiasing Prefilters

An anti-aliasing filter is a filter used before a signal sampler to restrict the bandwidth of a signal to satisfy the sampling theorem approximately or completely over the band of interest. However, aliasing occurs when signals are sampled too infrequently, giving the illusion of a lower frequency signal. Since most signals are not bandlimited, they must be made so by lowpass filtering before sampling.

To sample a signal at a desired rate fs and satisfy the conditions of the sampling theorem, the signal must be prefiltered by a lowpass analog filter, known as an antialiasing prefilter. The cutoff frequency of the prefilter, fmax, must be taken to be at most equal to the Nyquist frequency $f_s/2$, that is, $f_{max} \leq f_s/2$. This operation is shown in figure 3.5.

The output of the analog prefilter will then be bandlimited to maximum frequency fmax and may be sampled properly at the desired rate f_s. The spectrum replication caused by the sampling process can also be seen in figure 3.5.

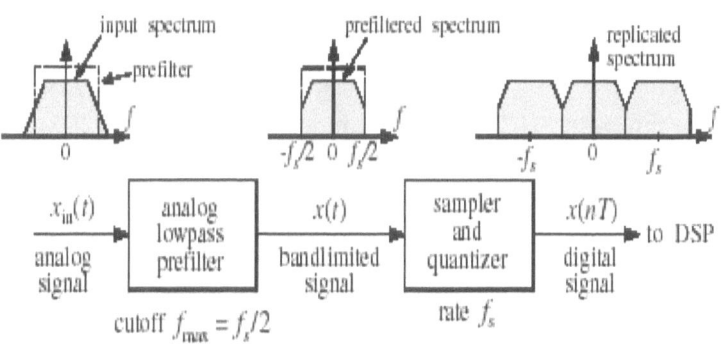

Figure 3.5. Antialiasing prefiter.

It should be emphasized that the rate fs must be chosen to be high enough so that, after the prefiltering operation, the surviving signal spectrum within the Nyquist interval

$[-f_s/2, f_s/2]$ contains all the significant frequency components for the application at hand.

Consider in a hi-fi digital audio application, we wish to digitize a music piece using a sampling rate of 40 kHz. Thus, the piece must be prefiltered to contain frequencies up to 20 kHz. After the prefiltering operation, the resulting spectrum of frequencies is more than adequate for this application because the human ear can hear frequencies only up to 20 kHz.

3.7 Types of Sampling

Sampling is the processes of converting continuous-time analog signal, $x_c(t)$, into a discrete-time signal by taking the "samples" at discrete-time intervals. Sampling analog signals makes them discrete in time but still continuous valued. A sample refers to a value or set of values at a point in time and/or space. A sampler is a subsystem or operation that extracts samples from a continuous signal.

A theoretical ideal sampler produces samples equivalent to the instantaneous value of the continuous signal at the desired points.

Typically, discrete-time signals are formed by periodically sampling a continuous-time signal: $x(n) = x_c(nT_s)$. The sampling interval T_s is the sampling period, and $f_s=1/T_s$ is the sampling frequency in samples per second. Figures 3.7, 3.8 show the sampling process.

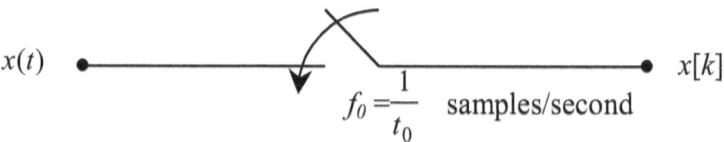

Figure 3.7 Periodic Sampling

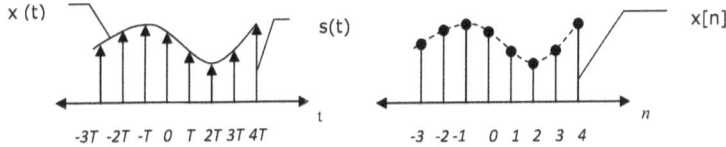

Figure 3.8. Sampling Process

There are three types of sampling techniques
- Impulse (ideal) Sampling
- Natural Sampling
- Sample and Hold operation

3.7.1 Impulse (Ideal) Sampling

Impulse sampling can be performed by multiplying input signal $x(t)$ with impulse train (comb function). Consider the instantaneous sampling of the analogue signal $x(t)$.

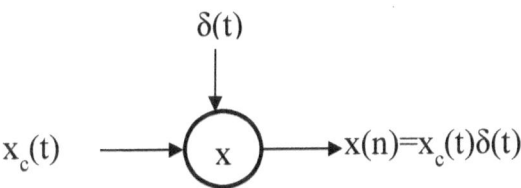

Figure 3.9 Ideal sampling

Impulse sampling can be performed by multiplying input signal $x(t)$ with impulse train $\sum_{n=-\infty}^{\infty} \delta(t - nT)$ of period 'T'. Here, the amplitude of impulse changes with respect to amplitude of input signal x(t). The output of sampler is given by

$$x_s(t) = x(t) \sum_{n=-\infty}^{\infty} \delta(t - nT_s)$$

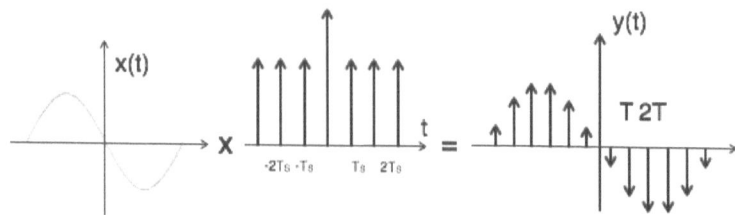

Figure 3.10. Ideal sampling process

$$y(t) = x(t) \times y(t) = x(t) \times impulse\ train$$
$$= x(t) \times \sum_{n=-\infty}^{\infty} \delta(t - nT)$$

$$y(t) = y_s(t) = \sum_{n=-\infty}^{\infty} \delta x(nt)(t - nT)$$

To get the spectrum of sampled signal, consider Fourier transform of equation 1 on both sides

$$Y(\omega) = \frac{1}{T} \sum_{n=-\infty}^{\infty} X(\omega - n\omega_0)$$

This is called ideal sampling or impulse sampling. You cannot use this practically because pulse width cannot be zero and the generation of impulse train is not possible practically.

3.7.2 Natural Sampling

Natural sampling is similar to impulse sampling, except the impulse train is replaced by pulse train of period T. i.e. you multiply input signal $x(t)$ to pulse train

$\sum_{n=-\infty}^{\infty} P(t - nT)$ as shown below.

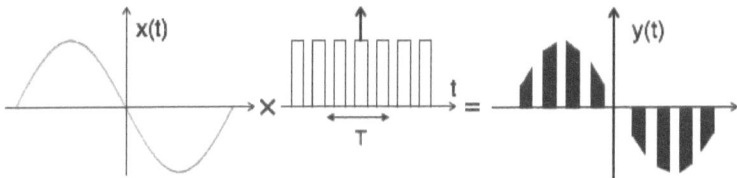

Figure 3.11. Natural sampling

The output of sampler is

$$y(t) = x(t) \times pulse\ train$$
$$= x(t) \times p(t)$$
$$= x(t) \times \sum_{n=-\infty}^{\infty} P(t - nT). \qquad (3.20) \qquad (3.20)$$

The exponential Fourier series representation of $P(t)$ can be given as

$$P(t) = \sum_{n=-\infty}^{\infty} F_n e^{jn\omega_0 t} \qquad (3.21)$$

Where $F_n = \frac{1}{T} \int_{\frac{-1}{T}}^{\frac{1}{T}} P(t) e^{-jn\omega_0 t}$

$$= \frac{1}{TP}(n\omega_0$$

Substitute F_n value in equation (3.20)

$$p(t) = \sum_{n=-\infty}^{\infty} \frac{1}{T} P(n\omega_s) e^{jn\omega_s t}$$

$$= \frac{1}{T} \sum_{n=-\infty}^{\infty} P(n\omega_s) e^{jn\omega_s t}$$

Substitute $P(t)$ in equation (3.21)

$$y(t) = x(t) \times p(t)$$

$$= x(t) \times \frac{1}{T}\sum_{n=-\infty}^{\infty} P(n\omega_s) e^{jn\omega_s t}$$

$$y(t) = \frac{1}{T}\sum_{n=-\infty}^{\infty} P(n\omega_s) x(t) e^{jn\omega_s t}$$

To get the spectrum of sampled signal, consider the Fourier transform on both sides.

$$F.T\left[y(t)\right] = F.T\left[\frac{1}{T}\sum_{n=-\infty}^{\infty} P(n\omega_s) x(t) e^{jn\omega_s t}\right]$$

$$= \frac{1}{T}\sum_{n=-\infty}^{\infty} P(n\omega_s) F.T\left[x(t) e^{jn\omega_s t}\right]$$

According to frequency shifting property

$$F.T\left[x(t) e^{jn\omega_s t}\right] = X[\omega - n\omega_s]$$

$$\therefore Y[\omega] = \frac{1}{T}\sum_{n=-\infty}^{\infty} P(n\omega_s) X[\omega - n\omega_s]$$

3.7.3 Sample-and-Hold (Flat Top) Sampling

Sample and hold is the most popular sampling method. it involves two operations: Sample and hold. During transmission, noise is introduced at top of the transmission pulse which can be easily removed if the pulse is in the form of flat top.

Here, the top of the samples are flat i.e. they have constant amplitude. Hence, it is called as flat top sampling or practical sampling. Flat top sampling makes use of sample and hold circuit.

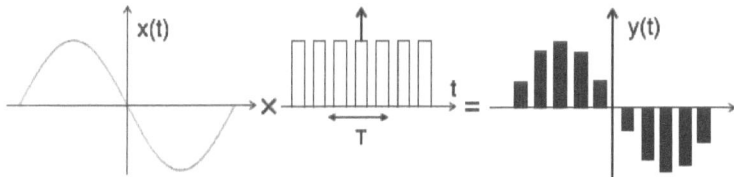

Figure 3.12. Flat Top sampling

Theoretically, the sampled signal can be obtained by convolution of rectangular pulse p(t) with ideally sampled signal say $y_s(t)$ as shown in figure 3.13.

$$y(t) = p(t) \times y_\delta(t) \tag{3.22}$$

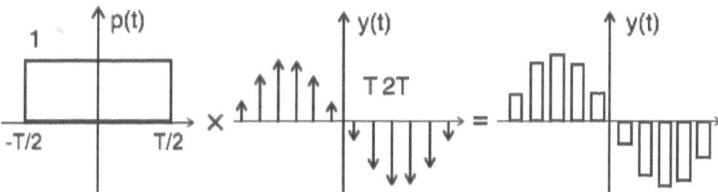

Figure 3.13. Sampled Signal

To get the sampled spectrum, consider Fourier transform on both sides for equation (3.20)

$$Y[\omega] = F.T\,[P(t) \times y_\delta(t)]$$

By the knowledge of convolution property,

$$Y[\omega] = P(\omega)\,Y_\delta(\omega)$$

Here

$$P(\omega) = TSa\left(\tfrac{\omega T}{2}\right) = 2\sin\omega T/\omega$$

Example 3.1
Consider the following signal

$$x(t) = 3\cos(100\pi t)$$

i. Find the minimum sampling rate required to avoid aliasing.
ii. If $F_s = 200$ Hz, What is the discrete-time signal after sampling?
iii. If $F_s = 75\ Hz$, What is the discrete-time signal after sampling?
iv. What is the frequency F of a sinusoidal that yields sampling identical to obtained in part iii?

Solution
i. Given, $\omega = 100\pi$, therefore, $F = 50$ Hz
The minimum sampling rate
is $F_s = 2F = 100$ Hz
and the discrete-time signal is
$$x(n) = x_c(nT) = 3\cos\frac{100\pi}{100}n = 3\cos\pi n = 3\cos(2\pi)\left(\tfrac{1}{2}\right)n$$

ii. If $F_s = 200$ Hz, the discrete-time signal is
$$x(n) = 3\cos\frac{100\pi}{200}n = 3\cos\frac{\pi}{2}n = 3\cos 2\pi\tfrac{1}{4}n$$

iii. If $F_s = 75$ Hz, the discrete-time signal is
$$x(n) = 3\cos\frac{100\pi}{75}n = 3\cos\frac{4\pi}{3}n = 3\cos\left(2\pi - \frac{2\pi}{3}\right)n = 3\cos 2\pi\tfrac{1}{3}n$$

iv. For the sampling rate, $F_s = 75$ Hz in part iii. So, the analog sinusoidal signal is
$$y_c(t) = 3\cos 2\pi Ft$$
$$= 3\cos 50\pi t$$

3.8 Quantization

It can be defined that the transformation of a signal $x[n]$ into one of a set of prescribed values. in fact it is objective is to make the signal amplitude discrete. The output of a sampler is still continuous in amplitude.

There are two types of Quantization - Uniform Quantization and Non-uniform Quantization. The type of quantization in which the quantization levels are uniformly spaced is termed as a Uniform Quantization. The type of quantization in which the quantization levels are unequal and mostly the relation between them is logarithmic, is termed as a Non-uniform Quantization. There are two types of uniform quantization. They are Mid-Rise type and Mid-Tread type. The following figures represent the two types of uniform quantization.

Figure 3.14(a) shows the mid-rise type and figure 3.14 (b) shows the mid-tread type of uniform quantization.

- The *Mid-Rise* type is so called because the origin lies in the middle of a raising part of the stair-case like graph. The quantization levels in this type are even in number.
- The *Mid-tread* type is so called because the origin lies in the middle of a tread of the stair-case like graph. The quantization levels in this type are odd in number.
- Both the mid-rise and mid-tread type of uniform quantizers are symmetric about the origin.

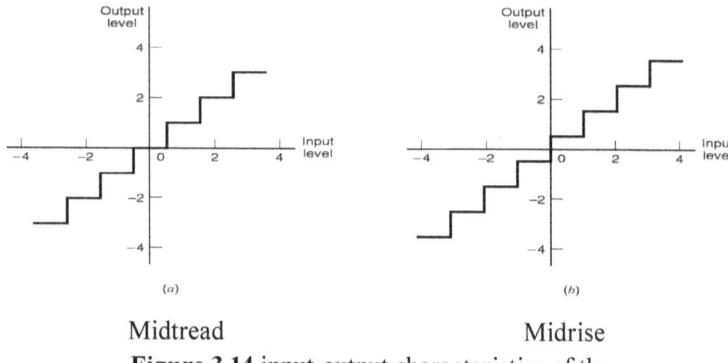

Midtread   Midrise

Figure 3.14 input-output characteristics of the two types of the uniform quantizers

Figure 3.14 shows the input-output characteristics of the two types of the uniform quantizer. In midtread type quantizer, the origin lies in the middle of the tread of the staircase like graph. In midrise type, the origin the graph lies in the middle of a rising part of the staircase like graph. It can be noticed that, both midtread and midrise graphs are symmetric about the origin.

3.8.1 Quantization Error

For any system, during its functioning, there is always a difference in the values of its input and output. The processing of the system results in an error, which is the difference of those values. The difference between an input value and its quantized value is called a Quantization Error as shown in figure 3.15.

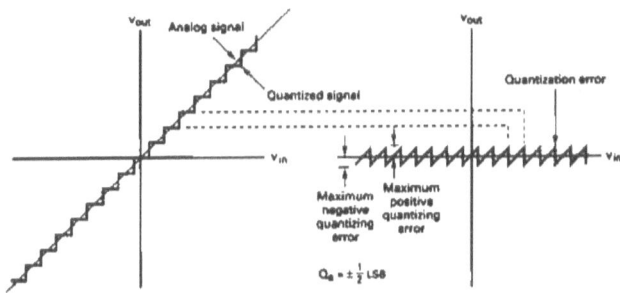

Figure 3.15 Quantization error

A Quantizer is a logarithmic function that performs Quantization rounding off the value rounding off the value. An analog-to-digital converter (ADC) works as a quantizer. When a signal is quantized, an error is introduced and some information is lost. The difference between the original sample value and the rounded value is called the quantization error. The process of quantization is to approximate to nearer level of voltage/current. Due to this approximation, there a random amount of difference occurs between actual and quantized value, named as quantization error as shown in figure 3.16.

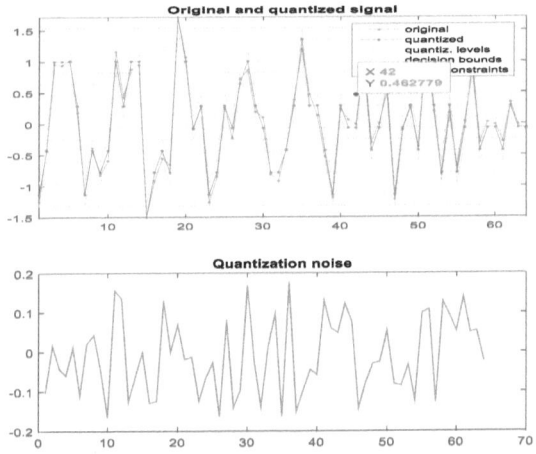

Figure 3.16 The difference between actual and quantized value, named as quantization error

The quantization operation introduces an error, because the infinite amount of different input amplitudes is mapped to a finite set of discrete quantization levels. The error between the quantizer's input and output is termed *Quantization noise* nq(t) nq(t). When denoting with s(t)s(t) the input signal and sq(t)sq(t) the output signal of the quantizer, the following relation holds:

Quantization operation
$$x_q(t) = Q[x(t)]$$
Expression of quantization noise
$$n_q(t) = x(t) - x_q(t)$$
$$x_q(t) = x(t) - n_q(t).$$

Looking at the last equation, the quantization noise can indeed be understood as a noise on top of the continuous-amplitude signal, hence its name. Let us create a sine wave and sent it through the quantizer and let's look at the quantized signal and the quantization noise.

signal to quantization noise ratio $SNR_q = S/N$ is given as:

$$(SNR)_q = \frac{Average\ Power\{x\}}{Average\ Power\{n_Q\}}$$

3.9 Ideal Reconstruction

Reconstruction refers using just the samples to return to the original continuous-time signal. Ideal reconstruction refers to exact reconstruction of from its samples so long as the sampling theorem is satisfied. In the extreme case example, this means that a sinusoid having frequency just less than, can be reconstructed from samples taken at rate. The block diagram of an ideal discrete-to-continuous (D-to C) converter is shown in figure 3.17.

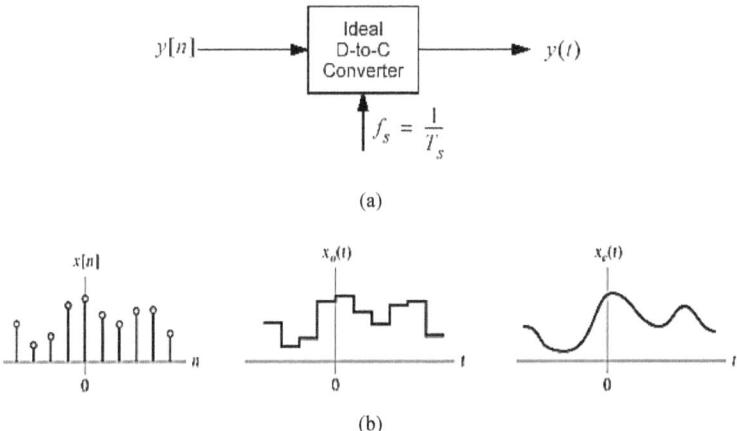

Figure 3.17 Block diagram of a practical reconstruction system

In very simple terms the D-to-C performs interpolation on the sample values $y[n]$ as they are placed on the time axis at spacing T_s second

Consider placing the sample values directly on the time axis as shown in figure 3.18.

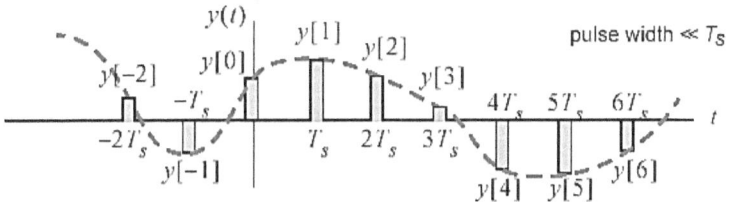

Figure 3.18

The D-to-C places the $y[n]$ values on the time axis and then must interpolate signal waveform values in between the sequence (sample) values. Two very simple interpolation functions are *zero-order hold* and *linear interpolation* With zero-order hold each sample value is represented as a rectangular pulse of width T_s and height $y[n]$.

Real world digital-to-analog converters (DACs) perform this type of interpolation. With linear interpolation the continuous waveform values between each sample value are formed by connecting a line between the $y[n]$ values as shown in figure 3.19.

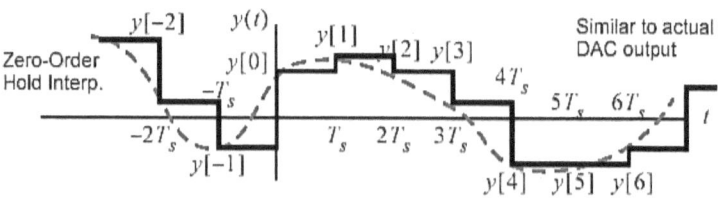

Figure 3.19

Both cases introduce errors, so it is clear that something better must exist

For D-to-C conversion using pulses, we can write

$$y(t) = \sum_{n=-\infty}^{\infty} y[n]p(t - nT_s)$$

where $P(t)$ is a rectangular pulse of duration T_s

Therefore, the complete sampling and reconstruction system requires both a C-to-D and a D-to-C

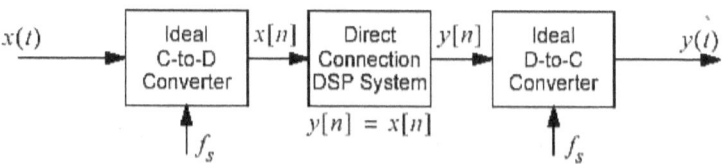

Figure 3.20 A complete sampling and reconstruction system

With this system we can sample analog signal $x(t)$ to produce $x[n]$, and at the very least we may pass $x[n]$ directly to $y[n]$, then reconstruct the samples $y[n]$ into $y(t)$.

3.10 Signal Reconstruction

The process of reconstructing a continuous time signal x(t) from its samples is known as interpolation. In the sampling theorem we saw that a signal x(t) band limited to D Hz can be reconstructed from its samples. This reconstruction is accomplished by passing the sampled signal through an ideal low pass filter of bandwidth D Hz

The reconstruction process consists of replacing each sample by a sinc function, centered at the time of the sample and scaled by the sample value $x(nT)$ times $2F_c/f_s$ and adding all the functions so created. Suppose the signal is sampled at exactly Nyquist rate $f_s = 2f_s$ Then $fm = f_s/2 = f_s - fm$ and $Fm = 1/2 = 1 - Fm$.

The requirement $Fm < Fc < 1 - Fm$ cannot be met, in this case we must allow $Fc = Fm$ which means that $fc = fm = 1/2$. . This will work till the signal's spectrum does not have an impulse at fm. (If there is an impulse at *fm*, it will be aliased in the sampling process). In this limiting case, the interppolation is described by the simpler expression.

If a sampled signal $x[n]$ has been obtained from a band-limited signal $x(t)$ by sampling at the Nyquist rate (or higher), the signal $x(t)$ can be perfectly reconstructed using the formula:

$$x(t) = \sum_{n=-\infty}^{\infty} x(nT) sinc(\frac{t-nT}{T}) \quad (3.23)$$

Interpolation consists of simply of multiplying each sinc function by its corresponding sample value and then adding all the scaled and shifted sinc functions.

However, this represents a filter that has an infinite impulse response, which is therefore non-causal.

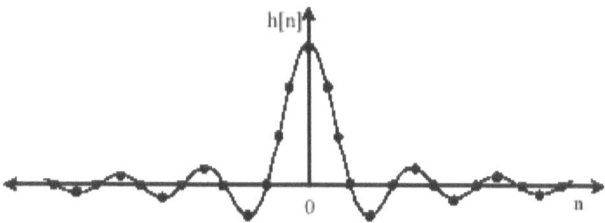

Figure 3.21 Impulse response of perfect reconstruction filter

In practice, therefore, it is usual to use the "zero-order hold "reconstruction filter which has the formula:

Problems

3.1 Consider a sinusoidal signal

$$x(t) = 3\cos(\pi t + 0.1\pi)$$

and let us sample it at a frequency $Fs = 2$ kHz

i. Determine and expression for the sampled sequence $x(n) = x(nT_s)$ and determine its Discrete Time Fourier Transform $X(\omega) = DTFT\{x[n]\}$.
ii. Determine $X(F) = FT\{x(t)\}$
iii. Recompute $X(\omega)$ from the $X(F)$ and verify that you obtain the same expression as in **i**)

3.2 In the system shown, let the sequence be
$y[n] = 2\cos(0.3\pi n + \pi/4)$.
and the sampling frequency be $Fs = 4$ kHz. Also let the low pass filter be ideal, with bandwidth $Fs/2$.

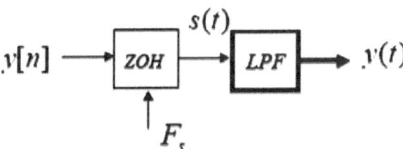

i. Determine an expression for $S(F) = FT\{s(t)\}$. Also sketch the frequency spectrum (magnitude only) within the frequency range $-Fs < F < Fs$;
ii. Determine the output signal $y(t)$.

3.3 We want to digitize and store a signal on a CD, and then reconstruct it at a later time. Let the signal $x(t)$ be

$$x(t) = 2\cos(500\pi t) - 3\sin(1000\pi t) + \cos(1500\pi t)$$

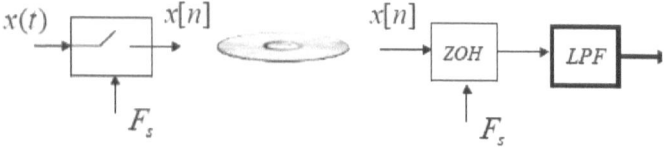

and let the sampling frequency be $F_s = 2000$ Hz.

i. Determine the continuous time signal $y(t)\wedge$ after the reconstruction.

ii. Notice that $y(t)$ is not exactly equal $x(t)$. How could we reconstruct the signal $x(t)$ exactly from its samples $x(n)$?

3.4 A DSP system for processing sinusoidal signals in the frequency range 0 Hz to 4 kHz samples at 20 kHz with an 8-bit A/D converter. If the input signal is always amplified to such a level that the full dynamic range of the A/D converter is used, estimate the SQNR that is to be expected in the frequency range 0 to 4 kHz. How would the SQNR be affected by decreasing the sampling rate to 10 kHz & replacing the 8-bit A/D converter by a 10 bit device?

3.5 In the system shown below, determine the output signal $y(t)$ for each of the following input signals $x(t)$. Assume the sampling frequency Fs = 5 kHz and the Low Pass Filter (LPF) to be ideal with bandwidth Fs / 2:

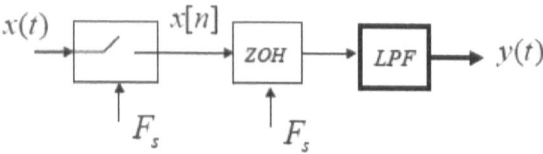

i. $x(t) = e^{j2000\pi t}$;
ii. $x(t) = \cos(2000\pi t + 0.15\pi)$
iii. $x(t) = 2\cos(5000\pi t)$;

iv. $x(t) = 2\sin(5000\pi t)$;
v. $x(t) = \cos(2000\pi t + 0.1\pi) - \cos(5500\pi t)$.

3.6 Instead of using sample and hold (staircase) digital to analogue reconstruction, a system reconstructs analogue output signals using pulses of the same height but only $T/2$ seconds wide & zero in between. This makes the pulses a little bit more like impulses and reduces the 'sample & hold' roll-off effect. Using pulses of width $T/4$ makes the pulses even more like impulses and reduces the sample & hold effect still further. Is this a good way of more accurately converting from digital to analogue form?

3.7 In the system below, let the sampling frequency be $Fs = 10$ kHz and the digital filter have difference equation

$$y[n] = 0.25(x[n] + x[n-1] + x[n-2] + x[n-3])$$

Both analog filters (Antialiasing and Reconstruction) are ideal Low Pass Filters (LPF) with bandwidth Fs / 2.

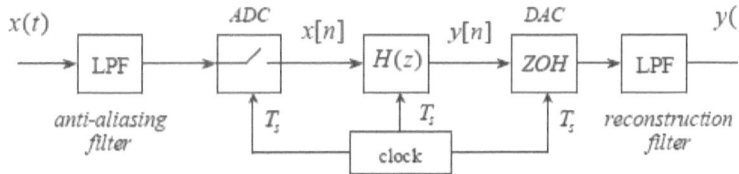

a. Sketch the frequency response $H(\omega)$ of the digital filter (magnitude only);
b. Sketch the overall frequency response $Y(F)/X(F)$ of the filter, in the analog domain (again magnitude only);
c. Let the input signal be
$x(t) = 3\cos(6000\pi t + 0.1\pi) - 2\cos(12000\pi t)$
d. Determine the output signal $y(t)$.

3.8 Assuming that a 4-bit ADC channel accepts analog input ranging from 0 to 5 volts, determine the following: a. number of quantization levels; b. step size of the quantizer or resolution; c. quantization level when the analog voltage is 3.2 volts; d. binary code produced by the ADC; e. quantization error.

3.9 Assuming that a 3-bit ADC channel accepts analog input ranging from 2:5 to 2.5 volts, determine the following:

   i. number of quantization levels;
   ii. step size of the quantizer
   iii. or resolution;
   iv. quantization level when the analog voltage is -1.2 volts
   v. binary code
   vi. produced by the ADC;
   vii. quantization error

3.10 If the analog signal to be quantized is a sinusoidal waveform, that is,

   i. $x(t) = 9.5 \sin(2000\pi t)$, and if the bipolar quantizer
   ii. uses 6 bits, determine
   iii. number of quantization levels;
   iv. quantization step size or resolution, Δ, assuming that the signal range is from -10 to 10 volts;
   v. the signal power to quantization noise power ratio.

Chapter Four

Discrete-Time Signals and Systems

Learning Outcomes of this Chapter

After successful completion of this chapter students will be able to:

1. understanding deterministic and random discrete-time signals and ability to generate them.
2. recognize the discrete-time system properties, namely, memorylessness, stability, causality, linearity and time-invariance.
3. understanding the relationship between difference equations and discrete-time signals and systems.

4.1 Discrete-Time Signals

A discrete-time (DT) signal is signal that exists at specific time instants. The amplitude of a discrete-time signal can be continuous though. When the amplitude of a DT signal is also discrete, then the signal is a digital signal. A DT signal can be either real or complex. While a real signal carries only amplitude information about a physical phenomenon, a complex signal carries both amplitude and phase information. A sequence of data is denoted $\{x[n]\}$ or simply $x[n]$ when the meaning is clear. The elements of the sequence are called samples. The index n associated with each sample is an integer. If appropriate, the range of n will be specified. Quite often, we are interested in identifying the sample where $n = 0$. This is done by putting an arrow under that sample. For instance,

$$\{x[n]\} = \{\ldots, 0.35, 1, 1.5, -0.6, -2, \ldots\}$$
$$\uparrow$$

The arrow is often omitted if it is clear from the context which sample is $x[0]$. Sample values can either be real or complex. In the rest of this book, the terms "discrete-time signals" and "sequences" are used interchangeably. The time interval between samples is not explicitly shown. It can be assumed to be normalized to 1 unit of time. So, the corresponding normalized sampling frequency is 1 Hz.

If the actual sampling interval is T seconds, then the sampling frequency is given by $f_s = \frac{1}{T}$

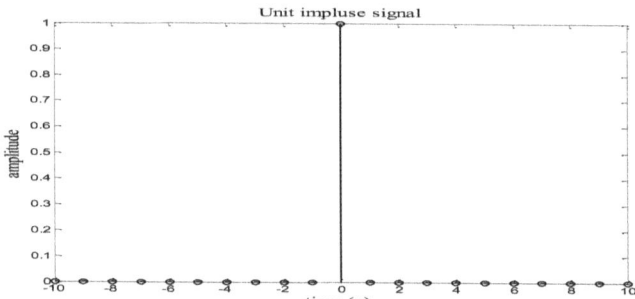

Figure 4.1 The Unit Impulse Sequence

4.1.1 Some Elementary Sequences

There are some sequences that we shall encounter frequently. They are described here.

4.1.1.1 Unit Impulse Sequence

The unit impulse sequence is defined by

$$\delta[n] = \begin{cases} 1, & n = 0 \\ 0, & n \neq 0 \end{cases} \tag{4.1}$$

This is depicted graphically in Figure 4.1. Note that while the continuous-time unit impulse function is a mathematical object that cannot be physically realized, the unit impulse sequence can easily be generated.

4.1.1.2 Unit Step Sequence

The unit step sequence is one that has an amplitude of zero for negative indices and an amplitude of one for non-negative indices. It is shown in Figure 4.2.

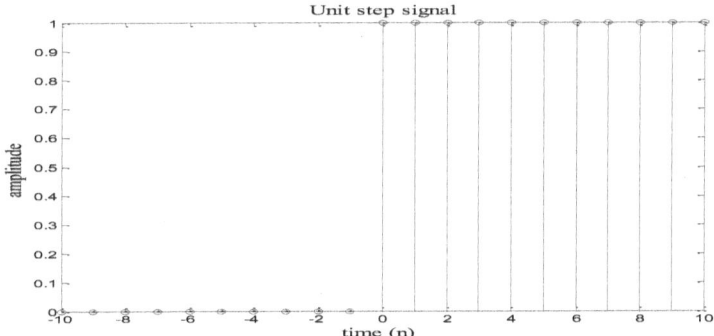

Figure 4.2 The Unit Step Sequence

4.1.1.3 The unit ramp signal

It denoted as $u_r(n)$ and is defined as follows:

$$u_r(n) \equiv \begin{cases} n, & \text{for } n \geq 0 \\ 0, & \text{for } n < 0 \end{cases}$$

It is shown in Figure 4.3.

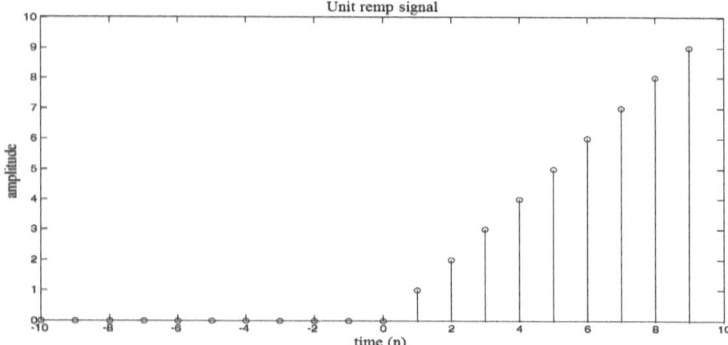

Figure 4.3 Unit ramp signal

4.1.1.4 Sinusoidal Sequences

A sinusoidal sequence has the form

$$x[n] = A\cos(\omega_0 n + \varphi) \qquad (4.2)$$

This function can also be decomposed into its in-phase $x_i[q]$ and quadrature $x_q[n]$ components.

$$x[n] = A\cos\varphi \cos\omega_0 n - A\sin\varphi \sin\omega_0 n \qquad (4.3)$$
$$= x_i[n] + x_q[n] \qquad (4.4)$$

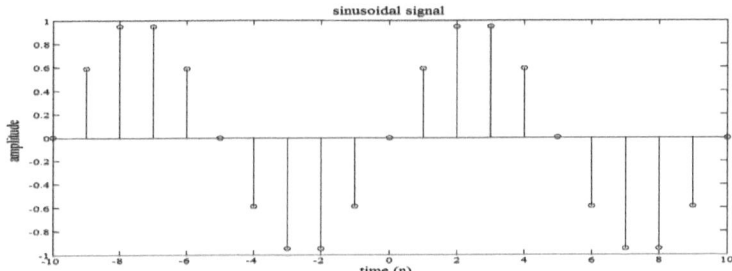

Figure 4.4 sinusoidal sequence

2.1.1.5 Complex Exponential Sequences

Complex exponential sequences are essentially complex sinusoids.

$$x[n] = Ae^{j(\omega_0 n + \varphi)} \tag{4.5}$$
$$= A\cos(\omega_0 n + \varphi) + jA\sin(\omega_0 n + \varphi)$$

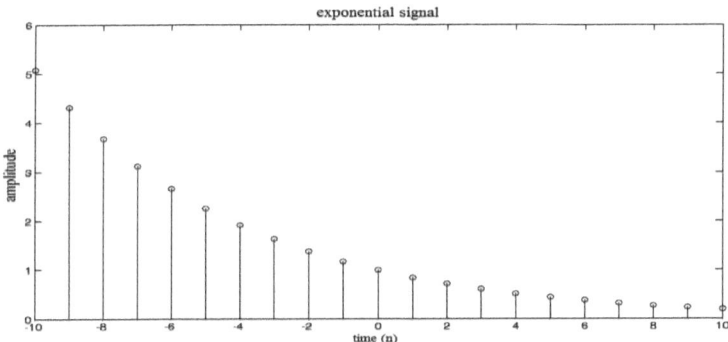

Figure 4.5 Exponential Sequences

4.1.1.6 Random Sequences

The sample values of a random sequence are randomly drawn from a certain probability distribution. They are also called stochastic sequences. The two most common distributions are the Gaussian (normal) distribution and the uniform distribution. The zero-mean Gaussian distribution is often used to model noise. Figure 4.6 and figure 4.7 show examples of uniformly distributed and Gaussian distributed random sequences respectively.

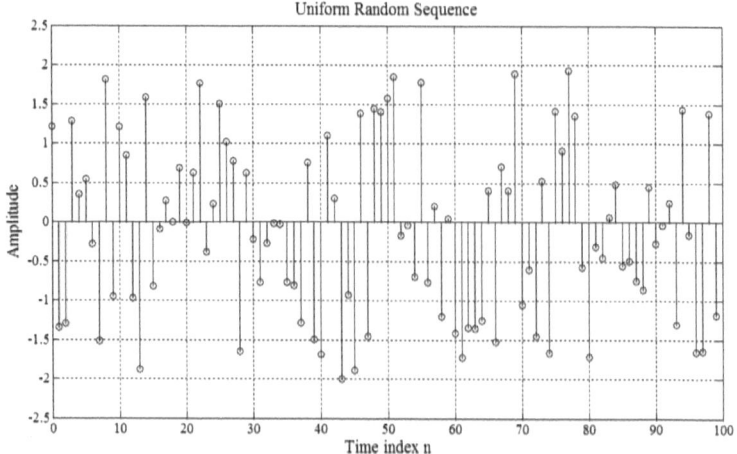

Figure 4.6 Random signal of length 100 with elements uniformly distributed

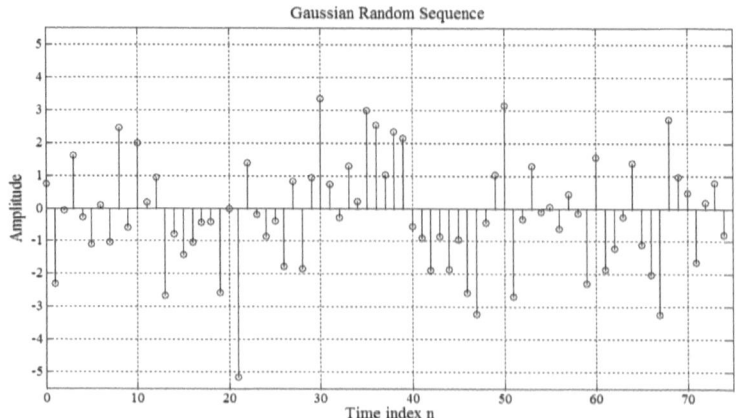

Figure 4.7 Gaussian distributed random sequence

4.2 Types of Sequences

The discrete-time signals that we encounter can be classified in several ways. Some basic classifications that are of interest to us are described below.

4.2.1 Real vs. Complex Signals

A sequence is considered complex at least one sample is complex-valued.

4.2.2 Finite vs. Infinite Length

Signals Finite length sequences are defined only for a range of indices, say N_1 to N_2. The length of this finite length sequence is given by $|N_2 - N_1 + 1|$.

4.2.3 Causal vs. Anti-casual Signals

A sequence $x[n]$ is a causal sequence if $x[n] = 0$ for $n < 0$.

Symmetric Signals First consider a real-valued sequence $\{x[n]\}$. Even symmetry implies that $x[n] = x[-n]$ and for odd symmetry $x[n] = -x[n]$ for all n. Any real-valued sequence can be decomposed into odd and even parts so that
$$x[n] = x_e[n] + x_o[n]$$

where the even part is given by
$$x_e[n] = \tfrac{1}{2}(x[n] + x[-n])$$

and the odd part is given by
$$x_o[n] = \tfrac{1}{2}(x[n] - x[-n])$$

A complex-valued sequence is conjugate symmetric if $x[n] = x^*[-n]$. The sequence has conjugate anti-symmetry if $x[n] = -x^*[-n]$. Analogous to real-valued sequences, any complex-valued sequence can be decomposed into its conjugate symmetric and conjugate anti-symmetric parts:

$$x[n] = x_{cs}[n] + x_{ca}[n] \tag{4.6}$$

$$x_{cs}[n] = \frac{1}{2}(x[n] + x^*[-n]) \tag{4.7}$$

$$x_{ca}[n] = \frac{1}{2}(x[n] - x^*[-n]) \tag{4.8}$$

Periodic Signals A discrete-time sequence is periodic with a period of N samples if

$$x[n] = x[n + kN] \tag{4.9}$$

for all integer values of k. Note that N has to be a positive integer. If a sequence is not periodic, it is aperiodic or non-periodic. We know that continuous-time sinusoids are periodic. For instance, the continuous-time signal

$$x(t) = cos(\omega_0 t) \tag{4.10}$$

has a frequency of ω0 radians per second or $f_0 = \omega_0/2\pi$ Hz. The period of this sinusoidal signal is $T = 1/f_0$ seconds.

Now consider a discrete-time sequence $x[n]$ based on a sinusoid with angular frequency ω0:

$$x[n] = cos(\omega_0 n) \tag{4.11}$$

If this sequence is periodic with a period of N samples, then the following must be true:

$$\cos(n+N) = \cos\omega_0 n \quad (4.12)$$

However, the left hand side can be expressed as

$$\cos\omega_0(n+N) = \cos(\omega_0 n + \omega_0 N) \quad (4.13)$$

and the cosine function is periodic with a period of 2π and therefore the right hand side of (4.13) is given by

$$\cos\omega_0 n = \cos(\omega_0 n + 2\pi r) \quad (4.14)$$

for integer values of r. Comparing (4.14) with (4.15), we have

$$\omega_0 N = 2\pi r$$
$$\rightarrow 2\pi f_0 N = 2\pi r$$
$$= \frac{r}{N} \quad (4.15)$$

where $\omega_0 = 2\pi f_0$. Since both r and N are integers, a discrete-time sinusoidal sequence is periodic if its frequency is a rational number. Otherwise, it is non-periodic.

Example 4.1. Is $x[n] = \cos(\frac{\pi}{8})n$ periodic? If so, what is the period? The sequence can be expressed as

$$x[n] = \cos(2\pi \left(\frac{1}{16}\right) n)$$

So in this case, $f_0 = 1/16$ is a rational number and the sinusoidal sequence is periodic with a period $N = 16$.

Example 4.2.
Determine the fundamental period of the following sequence:

$$x[n] = \cos(1.1\pi n) + \sin(0.7\pi n)$$

Solution
For the cosine function, the angular frequency is

$$\omega_1 = 1.1\pi = 2\pi(0.55) = 2\pi f_1$$

Therefore,

$$f_2 = \frac{55}{100} = \frac{11}{20}$$

and the period is $N_1 = 20$.
For the sine function, the angular frequency is
$$\omega_2 = 0.7\pi = 2\pi(0.35) = 2\pi f_2$$

where

$$f_2 = \frac{35}{100} = \frac{7}{20}$$

and the period is $N_2 = 20$.
So the period of $x[n]$ *is* 20.

It is interesting to note that for discrete-time sinusoidal sequences, a small change in frequency can lead to a large change in period. For example, a certain sequence has frequency $f_1 = 0.51 = 51/100$. So its period is 100 samples. Another sinusoidal sequence with frequency $f_2 = 0.5 = 50/100$ has a period of only 2 samples since f_2 can be simplified to 1/2. Thus a frequency difference of 0.01 can cause the period of the two sinusoidal sequences to differ by 98 samples.

Another important point to note is that discrete-time sinusoidal sequences with frequencies separated by an integer multiple of 2π are identical.

4.2.4 Energy and Power Signals

The energy of a finite length sequence $x[n]$ is defined as

$$E = \sum_{N_1}^{N_0}[x|n|]^2 \tag{4.17}$$

while that for an infinite sequence is

$$E = \sum_{-\infty}^{\infty}[x|n|]^2 \tag{4.18}$$

Note that the energy of an infinite length sequence may not be infinite. A signal with finite energy is usually referred to as an energy signal.

Example 4.3.
Find the energy of the infinite length sequence

$$x[n] = \begin{cases} 2^{-2n}, & n \geq 0 \\ 0, & n < 0 \end{cases}$$

According to the definition, the energy is given by

$$E = \sum_{n=0}^{\infty} 2^{-2n} = \sum_{n=0}^{\infty} \left(\frac{1}{n}\right)^n$$

To evaluate the finite sum, first consider

$$S_N = \sum_{n=0}^{\infty} 2^{-2n} = \sum_{n=0}^{\infty} \left(\frac{1}{n}\right)^n \tag{4.19}$$

Multiplying this equation by a, we have

$$aS_N = a + a^2 + \cdots \ldots \ldots + a^N \tag{4.20}$$

and the difference between these two equations give:

$$S_N - aS_N = (1 - a)S_N = 1 - a^N \tag{4.21}$$

Hence if $a \neq 1$

$$S_N = \frac{1-a^N}{1-a} \tag{4.22}$$

For $a = 1$, it is obvious that $S_N = N$. For $a < 1$, the infinite sum is therefore

$$S_\infty = \frac{1}{1-a} \tag{4.23}$$

Making use of this equation, the energy of the signal is

$$E = \frac{1}{1 - 1/4} = \frac{4}{3}$$

Equations (4.22) and (4.23) are very useful and we shall be making use of them later. The average power of a periodic sequence with a period of N samples is defined as

$$P_x = \frac{1}{N} \sum_{n=0}^{N-1} |x[n]|^2 \tag{4.24}$$

and for non-periodic sequences, it is defined in terms of the following limit if it exists:

$$P_x = \lim_{K \to \infty} \frac{1}{2K+1} \sum_{n=-K}^{K} |x[n]|^2 \tag{4.25}$$

A signal with finite average power is called a power signal.

Example 4.4
Find the average power of the unit step sequence $u[n]$.

The unit step sequence is non-periodic, therefore the average power is

$$P = \lim_{K \to \infty} \frac{1}{2K+1} \sum_{n=0}^{\infty} u^2[n]$$
$$= \lim_{K \to \infty} \frac{K+1}{2K+1}$$
$$= \frac{1}{2}$$

Therefore, the unit step sequence is a power signal. Note that its energy is infinite and so it is not an energy signal.

4.2.5. Bounded Signals: A sequence is bounded if every sample of the sequence has a magnitude which is less than or equal to a finite positive value. That is,

$$|x[n]| \leq B_x < \infty \qquad (4.26)$$

4.2.6. Summable Signals: A sequence is absolutely summable if the sum of the absolute value of all its samples is finite.

$$\sum_{n=-\infty}^{\infty} |x[n]| < \infty \qquad (4.27)$$

A sequence is square summable if the sum of the magnitude squared of all its samples is finite.

$$\sum_{n=-\infty}^{\infty} |x[n]|^2 < \infty \qquad (4.28)$$

4.3 Some Basic Operations on Sequences

Scaling: Scaling is the multiplication of a scalar constant with each and every sample value of the sequence. This operation is depicted schematically in Figure 4.8.

Addition: Addition of two sequences usually refers to point-by-point addition as shown in Figure 4.9.

Figure 4.8 Scalar Multiplication by A.

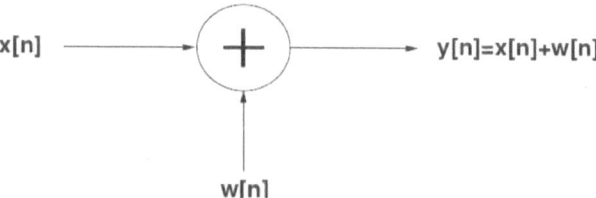

Figure 4.9 Point-by-point Addition

Delay: A unit delay shifts a sequence by one unit of time as shown in Figure 4.10. A sequence $x[n]$ is delayed by N samples to produce $y[n]$ if $y[n] = x[n - N]$.

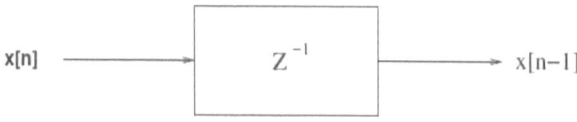

Figure 4.10 A Unit Delay

Up/Down Sampling: Down-sampling by a factor of L (a positive integer) is an operation by which only one every L-th sample of the original sequence is kept, with the rest discarded. The schematic for a down-sampler is shown in Figure 4.11 and 4.12. The down-sampled signal $y[n]$ is given by

$$y[n] = x[nM] \qquad (4.29)$$

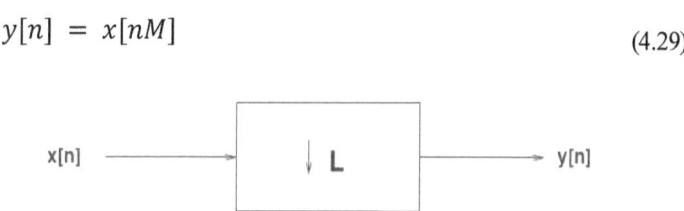

Figure 4.11 Down-sampling by a factor of L

Figure 4.12 Up-sample by a factor of L.

Up-sampling is the opposite operation to down-sampling. It increases the number of samples of the original sequence by a certain factor L (a positive integer). This is done by inserting $L - 1$ zeros between each pair of original samples.

The up-sampler and the up-sampled sequence y[n] is given by:

$$y[n] = \begin{cases} x\left[\frac{n}{L}\right], & n = 0, \pm L, \pm 2L, \ldots \\ 0, & Otherwise \end{cases} \qquad (4.30)$$

An interpolated signal can be obtained by passing the up-sampled signal through a lowpass filter with an appropriate bandwidth. This process will be discussed in more detail in a later chapter.

Modulation Given two sequences $x[n]$ and

$$w[n], \text{ and } y[n] = x[n] \cdot w[n] \qquad (4.31)$$

Then we say that $y[n]$ is $w[n]$ modulated by $x[n]$. This is analogous to carrier modulation in communication systems.

Correlation : The correlation, or more precisely cross-correlation, between two finite length data sequences $x[n]$ and $w[n]$ is defined by

$$r = \frac{1}{N} \sum_{n=0}^{N} x[n]w[n] \qquad (4.32)$$

if each sequence is of length N. The correlation coefficient r is often used as a measure of how similar the two sequences are. If they are very different, then the value of r is low.

The matched filter used in digital communication receivers for optimal detection is also effectively a correlator between the incoming and the template signals.

4.4 Discrete-time Systems

A discrete-time system is one that processes a discrete-time input sequence to produce a discrete-time output sequence. There are many different kinds of such systems.

4.4.1 Classification of Systems

Discrete-time systems, like continuous-time systems, can be classified in a variety of ways.

Linearity : A linear system is one which obeys the superposition principle. For a certain system, let the outputs corresponding to inputs $x_1[n]$ and $x_2[n]$ are $y_1[n]$ and $y_2[n]$ respectively. Now if the input is given by

$$x[n] = Ax_1[n] + Bx_2[n] \qquad (4.33)$$

where A and B are arbitrary constants, then the system is linear if its corresponding output is

$$y[n] = Ay_1[n] + By_2[n] \qquad (4.34)$$

Superposition is a very nice property which makes analysis much simpler. Although many real systems are not entirely linear throughout its operating region (for instance, the bipolar transistor), they can be considered approximately linear for certain input ranges. Linearization is a very useful approach to analyzing nonlinear systems. Almost all the discrete-time systems considered in this book are linear systems.

Example 4.5. Are the down-sampler and up-sampler linear systems? Consider the down-sampler

$$y[n] = x[nM]$$

For input x1[n], the corresponding output is $y_1[n] = x_1[nM]$. For input $x_2[n]$, the output is $y_2[n] = x_2[nM]$. Let x[n] be a linear combination of these two inputs with arbitrary constants A and B so that

$$x[n] = Ax_1[n] + Bx_2[n]$$

The output is given by
$$y[n] = Ax_1[nM] + Bx_2[nM]$$
$$= Ay_1[n] + By_2[n]$$

Therefore, the down-sampler is a linear system.
Now consider the up-sampler

$$y[n] = \begin{cases} x\left[\frac{n}{L}\right], & n = 0, \pm L, \pm 2L, \ldots \\ 0, & \text{Otherwise} \end{cases}$$

Let $y_1[n]$ and $y_2[n]$ be the outputs for inputs $x_1[n]$ and $x_2[n]$ respectively.

$$\text{For } x[n] = Ax_1[n] + Bx_2[n]$$

then the output is

$$y[n] = \begin{cases} Ax_1\left[\frac{n}{L}\right] + Bx_2\left[\frac{n}{L}\right], & n = 0, \pm L, \pm 2L, \ldots \\ 0, & \text{Otherwise} \end{cases}$$

$$= Ay_1[n] + By_2[n]$$

Hence the up-sampler is also linear.

Shift Invariance : A shift (or time) invariant system is one that does not change with time. Let a system response to an input $x[n]$ be $y[n]$. If the input is now shifted by n_0 (an integer) samples,

$$x_1[n] = x[n - n_0] \qquad (4.35)$$

then the system is shift invariant if its response to x1[n] is

$$y_1[n] = y[n - n_0] \qquad (4.36)$$

In the remainder of this book, we shall use the terms linear time-invariant (LTI) and linear shift-invariant interchangeably.

Example 4.6.
A system has input-output relationship given by

$$y[n] = \sum_{k=-\infty}^{n} x[k]$$

Is this system shift-invariant?

If the input is now $x_1[n] = x[n - n_0]$, then the corresponding output is

$$\begin{aligned}
y_1[n] &= \sum_{k=-\infty}^{n} x_1[k] \\
&= \sum_{k=-\infty}^{n} x[k - n_0] \\
&= \sum_{k=-\infty}^{n-n_0} x[k] \\
&= y[n - n_0]
\end{aligned}$$

Therefore, the system is shift invariant.

Example 4.7. Is the down-sampler a shift invariant system?

Let M (a positive integer) be the down-sampling ratio. So for an input $x[n]$ the output is

$$y[n] = x[nM].$$

Now if $x_1[n]$ is $x[n]$ delayed by n_0 samples, then

$$x_1[n] = x[n - n_0]$$

and the corresponding output is

$$y_1[n] = x[nM - n_0]$$
$$= x\left[\left(n - \frac{n_0}{M}\right)M\right]$$

If the system is shift invariant, one would expect the output to be

$$y[n - n_0] = x[(n - n_0)M]$$

Since this is not the case, the down-sampler must be shift variant.

Causality: The response of a causal system at any time depends only on the input at the current and past instants, not on any "future" samples. In other words, the output sample $y[n_0]$ for any n_0 only depends on $x[n]$ for $n \leq n_0$.

Example 4.8. Determine if the following system is causal:

$$y[n] = \sum_{k=-\infty}^{\infty} (n-k)u[n-k]x[k]$$

Note that $u[n - k] = 0$ for $n < k$ because the unit step sequence is zero for negative indices. In other words, for a certain n, $u[n - k] = 0$ for $k > n$. So the output can be written as

$$y[n] = \sum_{k=-\infty}^{\infty}(n-k)x[k]$$

So $y[n]$ depends on $x[k]$ for $k \leq n$ and therefore the system is causal.

Stability: There are two common criteria for system stability. They are exponential stability and bounded-input bounded-output (BIBO) stability. The first criterion is more stringent.

It requires the response of the system to decay exponentially fast for a finite duration input. The second one merely requires that the output be a bounded sequence if the input is a bounded sequence.

Example 4.9. Determine if the system with the following input-output relationship is BIBO stable.

$$y[n] = \sum_{k=-\infty}^{\infty} (n-k)u[n-k]x[k]$$

Consider input $x[n] = \delta[n]$.

Then

$$y[n] = \sum_{k=-\infty}^{\infty} (n-k)u[n-k]\delta[k]$$

$$= nu[n]$$

which is unbounded as it grows linearly with n. Therefore the system is not BIBO stable.

Example 4.10. Determine if the following system is BIBO stable. Note that this system is an "averager", taking the average of the past M samples.

$$y[n] = \frac{1}{M}\sum_{k=0}^{M-1} x(n-k)$$

Let the input $|x[n]| \leq B$ for some finite value B. Consider the magnitude of the output

$$|y[n]| = \left|\frac{1}{M}\sum_{k=0}^{M-1} x(n-k)\right|$$
$$\leq \frac{1}{M}\sum_{k=0}^{M-1} x(n-k)$$
$$\leq \frac{1}{M}(MB)$$
$$= B$$

Hence the output is bounded and the system is BIBO stable.

Lossy or Lossless : For a passive system that does not generate any energy internally, the output should have at most the same energy as the input. So

$$\sum_{n=-\infty}^{\infty}|y[n]|^2 \leq \sum_{n=-\infty}^{\infty}|x[n]|^2 < \infty \tag{4.37}$$

A lossless system is one which the equality holds.

4.4.2 Linear Shift-Invariant Systems

An discrete-time LTI system, like its continuous-time counterpart, is completely characterized by its impulse response. In other words, the impulse response tells us everything we need to know about an LTI system as far as signal processing is concerned. The impulse response is simply the observed system output when the input is an impulse sequence. For continuous-time systems, the impulse function is purely a mathematical entity. However, for discrete-time systems, since we are dealing with sequences of numbers, the impulse sequence can realistically (and easily) be generated.

4.4.3 Linear Convolution

Let us consider a discrete-time LTI system with impulse response $h[n]$ as shown in Figure 4.13. What would be the output $y[n]$ of the system if the input $x[n]$ is as shown in Figure 4.14?

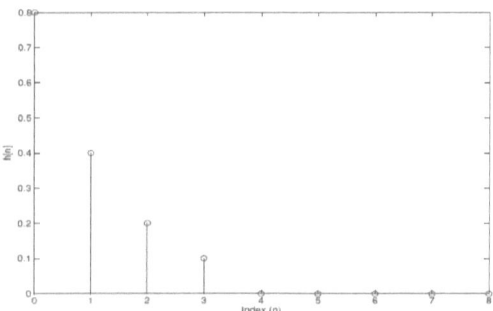

Figure 4.13: Impulse Response of the System

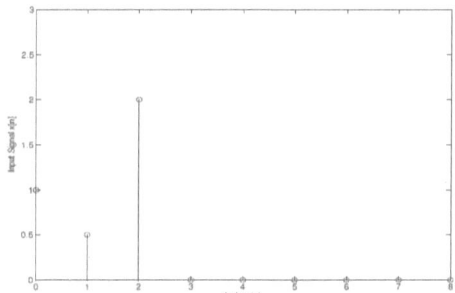

Figure 4.14 Input Signal Sequence

Since the system is linear and time invariant, we can make use of the superposition principle to compute the output. The input sequence is composed of three impulses. Mathematically, it can be expressed as

$$x[n] = \delta[n] + 0.5\delta[n-1] + 2\delta[n-2] \quad (4.38)$$

Let $x1[n] = \delta[n]$
$x2[n] = 0.5\delta[n-1]$
$x3[n] = 2\delta[n-2]$

and the system response to each of these inputs are respectively $y1[n], y2[n]$ and $y3[n]$.

The sample values of $y1[n]$ are given by

$y1[0] = h[0]x1[0] = 0.8$
$y1[1] = h[1]x1[0] = 0.4$
$y1[2] = h[2]x1[0] = 0.2$
$y1[3] = h[3]x1[0] = 0.1$

which is the same as the impulse response since $x1[n]$ is a unit impulse. Similarly,

$y2[1] = h[0]x2[1] = 0.4$
$y2[2] = h[1]x2[1] = 0.2$
$y2[3] = h[2]x2[1] = 0.1$
$y2[4] = h[3]x2[1] = 0.05$

and

$y3[2] = h[0]x3[2] = 1.6$
$y3[3] = h[1]x3[2] = 0.8$
$y3[4] = h[2]x3[2] = 0.4$
$y3[5] = h[3]x3[2] = 0.2$ The system output $y[n]$ in response to input $x[n]$ is therefore, through the superposition principle, given by

$$y[n] = y1[n] + y2[n] + y3[n]$$
$$= \{0.8, 0.8, 2, 1, 0.45, 0.2\}$$

Note that

$$y[0] = h[0]x[0]$$
$$y[1] = h[1]x[0] + h[0]x[1]$$
$$y[2] = h[2]x[0] + h[1]x[1] + h[0]x[2]$$
$$y[3] = h[3]x[0] + h[2]x[1] + h[1]x[2]$$
$$y[4] = h[3]x[1] + h[2]x[2]$$
$$y[5] = h[3]x[2]$$

In general, we have

$$y[n] = h[n]x[0] + h[n-1]x[1] + \ldots + h[1]x[n-1] + h[0]x[n] \quad (2.39)$$

or

$$y[n] = \sum_{k=0}^{n} h[k]x[n-k] \quad (4.40)$$

Alternatively,

$$y[n] = \sum_{k=0}^{n} x[k]h[n-k] \quad (4.41)$$

Equations 2.40 and 2.41 are the linear convolution equations for finite length sequences. If the length of $h[n]$ is M and the length of $x[n]$ is N, then the length of $y[n]$ is $N + M - 1$.

We can further generalize it for infinite length sequences:

$$y[n] = \sum_{k=-\infty}^{\infty} h[k]x[n-k] \quad (4.42)$$
$$= \sum_{k=-\infty}^{\infty} x[k]h[n-k] \quad (4.43)$$

These equations are analogous to the linear convolution equation for continuous-time signals. Note that the linear convolution equation comes about because of the superposition principles and therefore applies only to LTI systems.

The convolution equation for discrete-time signals is also called the convolution sum. It is denoted by ?. So equations 4.42 and 4.43 can be written as

$$y[n] = x[n] * h[n] \tag{4.44}$$

The convolution sum is one of the most important fundamental equations in DSP.

4.4.4 Properties of Linear Convolution

The convolution sum has three important properties:

1. Commutative
$$x[n] \; y*[n] = y[n] * x[n] \tag{4.45}$$

2. Associative
$$(x[n] * w[n]) * y[n] = x[n] * (w[n] * y[n]) \tag{2.46}$$

3. Distributive
$$x[n] * (w[n] + y[n]) = x[n] * w[n] + x[n] * y[n] \tag{2.47}$$

4.4.4.1 Condition for Stability

Since the impulse response completely characterize an LTI system, we should be able to draw conclusions regarding the stability of a system based on its impulse response. We shall consider BIBO stability here.

Theorem 4.1. A discrete-time LTI system is BIBO stable if its impulse response is absolutely summable.

Proof. Let the input be bounded, i.e. $|x[n]| < B < \infty$ for some finite value B. The magnitude of the output is given by

$$|y[n]| = \left| \sum_{k=-\infty}^{\infty} h[k]x[n-k] \right|$$
$$\leq \sum_{k=-\infty}^{\infty} h[k]x[n-k]$$
$$\leq B \sum_{k=-\infty}^{\infty} |h[k]|$$

So, the magnitude of $y[n]$ is bounded if $\sum_{k=-\infty}^{\infty} |h[k]|$ is finite. In other words, the impulse response must be absolutely summable.

4.4.4.2 Condition for Causality

Theorem 4.2. A discrete-time LTI system is causal if and only if its impulse response is a causal sequence.

Proof. Consider an LTI system with impulse response $h[k]$. Two different inputs $x_1[n]$ and $x_2[n]$ are the same up to a certain point in time, that is $x_1[n] = x_2[n]$ for $n \leq n_0$ for some n_0. The outputs $y_1[n]$ and $y_2[n]$ at $n = n_0$ are given by

$$y_1[n_0] = \sum_{k=-\infty}^{\infty} h[k]x_1[n_0 - k]$$
$$= \sum_{k=-\infty}^{-1} h[k]x_1[n_0 - k] + \sum_{k=0}^{\infty} h[k]x_1[n_0 - k]$$

and

$$y_2[n_0] = \sum_{k=-\infty}^{\infty} h[k]x_2[n_0 - k]$$

$$= \sum_{k=-\infty}^{-1} h[k]x_2[n_0 - k] + \sum_{k=0}^{\infty} h[k]x_2[n_0 - k]$$

Since $x_1[n] = x_2[n]$ for $n \leq n_0$, if the system is causal, then the outputs $y_1[n]$ and $y_2[n]$ must be the same for $n \leq n_0$. More specifically, $y_1[n_0] = y_2[n_0]$. Now,

$$\sum_{k=0}^{\infty} h[k]x_1[n_0 - k] = \sum_{k=0}^{\infty} h[k]x_2[n_0 - k]$$

because $x_1[n_0 - k] = x_2[n_0 - k]$ for non-negative values of k. Since $x_1[n]$ may not be equal to $x_2[n]$ for $n > n_0$, we must have

$$\sum_{k=-\infty}^{-1} h[k]x_1[n_0 - k] = \sum_{k=-\infty}^{-1} h[k]x_2[n_0 - k] = 0$$

which means that $h[k] = 0$ for $k < 0$.

Problems

4.1 Determine if the following signals are periodic, and if so compute the fundamental period.

(a) $x[n] = e^{j\frac{20\pi}{3}n}$

(b) $x[n] = e^{j4\pi n} - e^{-j\frac{\pi}{4}n}$

(c) $x[n] = 3e^{j\frac{7}{3}n}$

(d) $x[n] = 1 + e^{j\frac{4}{5}\pi n}$

(e) $x[n] = e^{j\frac{5}{7}\pi n} + e^{-j\frac{3}{4}\pi n}$

4.2 Simplify the following expressions and sketch the signal.

(a) $x[n] = \sum_{k=-\infty}^{n} 3\delta[k-2] + \delta[n+1]\cos(\pi n)$

(b) $x[n] = \sum_{k=-\infty}^{\infty} 4\delta[n-k]e^{3k}$

(c) $x[n] = r[n-3]\delta[n-5] + \sum_{k=-\infty}^{n} 3\delta[n-k]$

(d) $x[n] = \cos(\pi n)\big[\delta[n] - \delta[n-1]\big] - \delta^3[n] + \sum_{k=-\infty}^{n} u[k-3]$

4.3 Determine if each of the following systems is causal, memoryless, time invariant, linear, or stable. Justify your answer.

(a) $y[n] = 3x[n]x[n-1]$

(b) $y[n] = \sum_{k=n-2}^{n+2} x[k]$

(c) $y[n] = 4x[3n-2]$

(d) $y[n] = \sum_{k=-\infty}^{n} e^k u[k] x[n-k]$

(e) $y[n] = \sum_{k=n-3}^{n} \cos(x[k])$

4.4 Determine if each of the following systems is invertible. If not, specify two different input signals that yield the same output. If so, give an expression for the inverse system.

(a) $y[n] = \sum_{k=-\infty}^{n} x[k]$

(b) $y[n] = (n-1)x[n]$

(c) $y[n] = x[n] - x[n-1]$

4.5 Suppose an LTI system with input signal $x[n] = u[n] - u[n-2]$ has the response $y[n] = 2r[n] - 2r[n-2]$. Sketch this input signal and output signal, and also sketch the system response to each of the input signals below.

(a) $x_a[n] = 3u[n-1] - 3u[n-3]$
(b) $x_b[n] = u[n] - u[n-1] - u[n-2] + u[n-3]$
(c) $x_c[n] = u[n] - u[n-4]$

4.6 Using the graphical method, compute and sketch $y[n] = (h * x)[n]$ for

(a) $x[n] = \delta[n] - \delta[n-3]$, $h[n] = 3\delta[n+1] - 3\delta[n-3]$

(b) $x[n] = \begin{cases} 1, & 0 \le n \le 3 \\ 0, & else \end{cases}$, $h[n] = \begin{cases} 1, & 1 \le n \le 3, \ 7 \le n \le 9 \\ 0, & else \end{cases}$

(c) $x[n] = 1$, for all n, $h[n] = \delta[n] - 2\delta[n-1] + \delta[n-2]$
(d) $x[n] = u[n-1] - u[n-3]$, $h[n] = -u[n] + u[n-3]$
(e) $x[n] = e^n(u[n] - u[n-2])$, $h[n] = e^{-n}u[n]$
(f) $x[n] = u[n]$, $h[n] = (1/2)^n u[n-1]$
(g) $x[n] = r[n]$, $h[n]$ shown below

4.7 Determine if the DT LTI system with the following unit-pulse responses are causal and/or stable.

(a) $h[n] = (\frac{1}{2})^n u[n-1]$

(b) $h[n] = (\frac{1}{2})^n u[-n]$

(c) $h[n] = 2^n u[3-n]$

(d) $h[n] = 2^n r[-n]$

Chapter Five

Z-transform and applications

Learning Outcomes of this Chapter

After successful completion of this chapter students will be able to:

1. describe the difference between Fourier, Laplace, and z-transforms.
2. explain the relationship between z-transform pairs.
3. identify overall system behavior from a pole-zero diagram.
4. compute transform and inverse transform.
5. apply transform for analyzing linear time invariant (LTI) systems.

5.1 Introduction

Z-transform plays a similar role in DSP as the Laplace transform in analog circuits and systems. It is useful for the manipulation of discrete data sequences and has acquired a new significance in the formulation and analysis of discrete-time systems. This mathematical technique date back to the early 1730s when DeMoivre introduced the concept of a generating function that is identical with that for the Z-transform. Recently, the development and extensive applications of the Z-transform are much enhanced as a result of the use of digital computers It is used extensively today in the areas of applied mathematics, digital signalprocessing, control theory, population science, economics. It is used to define transfer functions and determine responses of systems using a table look-up process. The role played by the z-transform in the solution of difference equations corresponds to that played by the Laplace transforms in the solution of differential equations.

In mathematics and signal processing, the Z-transform (ZT) converts a discrete time-domain signal, which is a sequence of real or complex numbers, into a complex frequency-domain representation.

The ZT is a transformation that maps DT signal $x[n]$ into a function of the complex variable z, defined as:

$$X(z) = \sum_{n=-\infty}^{\infty} x[n]z^{-n} \qquad (5.1)$$

The z-transform can also be thought of as an operator $Z\{\cdot\}$ that transforms a sequence to a function:

$$\{x[n]\} = \sum_{n=-\infty}^{\infty} x[n]z^{-n} = X(z) \qquad (5.2)$$

In both cases z is a continuous complex variable.

We may obtain the Fourier transform from the z-transform by making the substitution $z = re^{j\omega}$. This corresponds to restricting $|z| = 1$. Also, with $z = re^{j\omega}$

$$X(z) = \sum_{n=-\infty}^{\infty} x[n](re^{j\omega})^{-n} = \sum_{n=-\infty}^{\infty} (x[n]r^{-n})re^{-j\omega n}$$

That is, the z-transform is the Fourier transform of the sequence $x[n]\, r^{-n}$ For $r = 1$ this becomes the Fourier transform of $x[n]$. The Fourier transform corresponds to the z-transform evaluated on the unit circle:

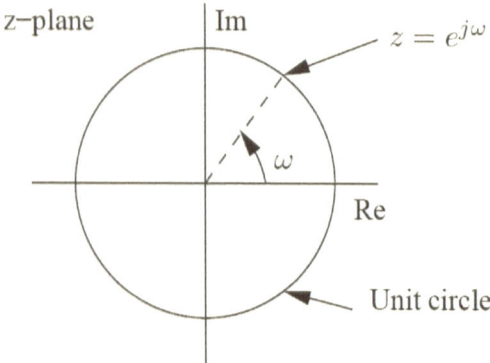

Figure 5.1 Z-Plan

The inherent periodicity in frequency of the Fourier transform is captured naturally under this interpretation. The Fourier transform does not converge for all sequences—the infinite sum may not always be finite. Similarly, the z-transform does not converge for all sequences or for all values of z. The set of values of z for which the z-transform converges is called the **region of convergence (ROC)**.

The Fourier transform of x[n] exists if the sum $\sum_{n=-\infty}^{\infty}|x[n[|$ converges. However, the z-transform of x[n] is just the Fourier transform of the sequence $x[n]\, r^{-n}$. The z-transform therefore exists (or converges) if

$$X(z) = \sum_{n=-\infty}^{\infty}|x[n]|r^{-n} < \infty \tag{5.3}$$

This leads to the condition

$$X(z) = \sum_{n=-\infty}^{\infty}|x[n]||z|^{-n} < \infty \tag{5.4}$$

for the existence of the z-transform. The ROC therefore consists of a ring in the z-plane:

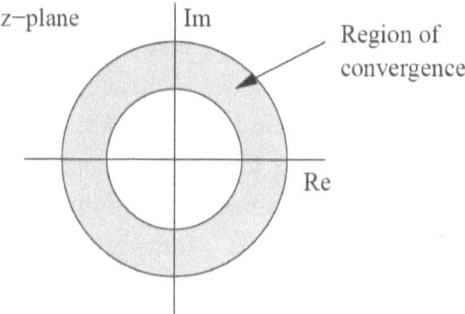

Figure 5.2 Region of Convergence

In specific cases the inner radius of this ring may include the origin, and the outer radius may extend to infinity. If the ROC includes the unit circle $|z| = 1$, then the Fourier transform will converge. Most useful z-transforms can be expressed in the form:

$$X(z) = \frac{N(z)}{D(z)}$$

where $N(n)$ and $D(z)$ are polynomials in z. The values of z for which $N(z) = 0$ are called the **zeros** of $X(z)$, and the values with $D(z) = 0$ are called the **poles**. The zeros and poles completely specify $X(z)$ to within a multiplicative constant.

5.2 Unilateral Z-transform

In cases where $x[n]$ is defined only for $n \geq 0$, the *single-sided* or *unilateral* Z-transform is defined as:

$$X(z) = Z\{x[n]\} = \sum_{n=0}^{\infty} x[n] z^{-n} \tag{5.5}$$

This is a z-transform of a causal sequence $x(n)$,

where z is the complex variable. Here, the summation taken from $n = 0$ to $n = 1$ is according to the fact that for most situations, the digital signal $x(n)$ is the causal sequence, that is, $x(n) = 0$ for $n < 0$. Thus, the definition in Equation (5.5) is referred to as a one-sided z-transform or a unilateral transform. In Equation (5.5), all the values of z that make the summation to exist form a region of convergence in the z-transform domain, while all other values of z outside the region of convergence will cause the summation to diverge. The region of convergence is defined based on the particular sequence $x(n)$ being applied. Note that we deal with the unilateral z-transform in this book, and hence when performing inverse z-transform (which we shall study later), we are restricted to the causal sequence. Now let us study the following typical examples.

Example 5.1.
Given the sequence $x(n) = u(n)$,
Find the z-transform of x(n).

Solution:
From the definition of Equation (5.5), the z-transform is given by

$$X(z) = \sum_{n=0}^{\infty} u[n]z^{-n} = \sum_{n=0}^{\infty} (z^{-1})^n = 1 + (z^{-1}) + (z^{-1})^2 + \ldots$$

This is an infinite geometric series that converges to :

$$X(z) = \frac{z}{z-1}$$

With a conclusion $|z^{-1}| < 1$. Note that for an infinite geometric series, we have $1 + r + r^2 + \ldots = \frac{1}{1-r}$ when $|r| < 1$. The region of convergence for all values of z is given as $|r| > 1$.

Example 5.2.
Considering the exponential sequence

$$x(n) = a^n u(n),$$

Find the z-transform of the sequence x(n).

Solution:
From the definition of the z-transform in Equation (5.5), it follows that

$$X(z) = \sum_{n=0}^{\infty} a^n u(n) z^{-n} = \sum_{n=0}^{\infty} (az^{-1})^n = 1 + (az^{-1}) + (az^{-1})^2 + \ldots$$

Since this is a geometric series which will converge for $|az^{-1}| < 1$, it is further expressed as

$$X(z) = \frac{z}{z-a}, \text{ for } |z| > |a|.$$

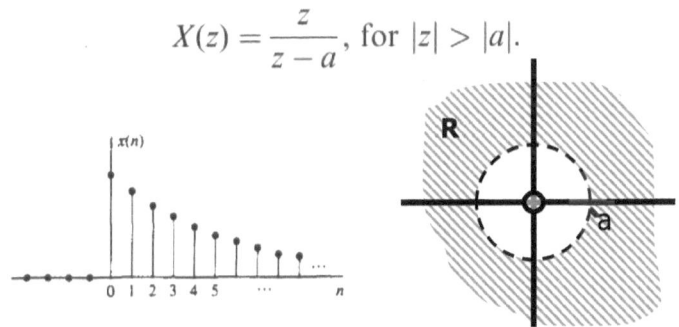

Figure 5.3 Shows the function in Example 5.2 and its ROC

The z-transforms for common sequences are summarized in Table 5.1.

Table 5.1 Table of z-transform pairs.

	Signal, $x[n]$	Z-transform, $X(z)$	ROC
1	$\delta[n]$	1	all z
2	$\delta[n - n_0]$	z^{-n_0}	$z \neq 0$
3	$u[n]$	$\dfrac{1}{1 - z^{-1}}$	$|z| > 1$
4	$e^{-\alpha n} u[n]$	$\dfrac{1}{1 - e^{-\alpha} z^{-1}}$	$|z| > |e^{-\alpha}|$
5	$-u[-n - 1]$	$\dfrac{1}{1 - z^{-1}}$	$|z| < 1$
6	$n u[n]$	$\dfrac{z^{-1}}{(1 - z^{-1})^2}$	$|z| > 1$
7	$-n u[-n - 1]$	$\dfrac{z^{-1}}{(1 - z^{-1})^2}$	$|z| < 1$
8	$n^2 u[n]$	$\dfrac{z^{-1}(1 + z^{-1})}{(1 - z^{-1})^3}$	$|z| > 1$
9	$-n^2 u[-n - 1]$	$\dfrac{z^{-1}(1 + z^{-1})}{(1 - z^{-1})^3}$	$|z| < 1$
10	$n^3 u[n]$	$\dfrac{z^{-1}(1 + 4z^{-1} + z^{-2})}{(1 - z^{-1})^4}$	$|z| > 1$
11	$-n^3 u[-n - 1]$	$\dfrac{z^{-1}(1 + 4z^{-1} + z^{-2})}{(1 - z^{-1})^4}$	$|z| < 1$
12	$a^n u[n]$	$\dfrac{1}{1 - a z^{-1}}$	$|z| > |a|$
13	$-a^n u[-n - 1]$	$\dfrac{1}{1 - a z^{-1}}$	$|z| < |a|$
14	$n a^n u[n]$	$\dfrac{a z^{-1}}{(1 - a z^{-1})^2}$	$|z| > |a|$
15	$-n a^n u[-n - 1]$	$\dfrac{a z^{-1}}{(1 - a z^{-1})^2}$	$|z| < |a|$
16	$n^2 a^n u[n]$	$\dfrac{a z^{-1}(1 + a z^{-1})}{(1 - a z^{-1})^3}$	$|z| > |a|$
17	$-n^2 a^n u[-n - 1]$	$\dfrac{a z^{-1}(1 + a z^{-1})}{(1 - a z^{-1})^3}$	$|z| < |a|$
18	$\cos(\omega_0 n) u[n]$	$\dfrac{1 - z^{-1} \cos(\omega_0)}{1 - 2z^{-1} \cos(\omega_0) + z^{-2}}$	$|z| > 1$
19	$\sin(\omega_0 n) u[n]$	$\dfrac{z^{-1} \sin(\omega_0)}{1 - 2z^{-1} \cos(\omega_0) + z^{-2}}$	$|z| > 1$
20	$a^n \cos(\omega_0 n) u[n]$	$\dfrac{1 - a z^{-1} \cos(\omega_0)}{1 - 2a z^{-1} \cos(\omega_0) + a^2 z^{-2}}$	$|z| > |a|$
21	$a^n \sin(\omega_0 n) u[n]$	$\dfrac{a z^{-1} \sin(\omega_0)}{1 - 2a z^{-1} \cos(\omega_0) + a^2 z^{-2}}$	$|z| > |a|$

We list the following table of z-transforms. It can also be used to find the inverse z-transform.

Example 5.3.

Find the z-transform for each of the following sequences:

a. $x(n) = 10u(n)$
b. $x(n) = 10\sin(0.25\pi n)u(n)$
c. $x(n) = (0.5)^n u(n)$
d. $x(n) = (0.5)^n \sin(0.25\pi n)u(n)$
e. $x(n) = e^{-0.1n}\cos(0.25\pi n)u(n)$

Solution:

a. From Table 5.1, we get

$$X(z) = Z(10u(n)) = \frac{10z}{z-1}.$$

b. Line 9 in Table 5.1 leads to

$$X(z) = 10Z(\sin(0.2\pi n)u(n))$$

$$= \frac{10\sin(0.25\pi)z}{z^2 - 2z\cos(0.25\pi) + 1} = \frac{7.07z}{z^2 - 1.414z + 1}.$$

c. From Line 6 in Table 5.1, we yield

$$X(z) = Z((0.5)^n u(n)) = \frac{z}{z - 0.5}.$$

d. From Line 11 in Table 5.1, it follows that

$$X(z) = Z((0.5)^n \sin(0.25\pi n)u(n)) = \frac{0.5 \times \sin(0.25\pi)z}{z^2 - 2 \times 0.5\cos(0.25\pi)z + 0.5^2}$$

$$= \frac{0.3536z}{z^2 - 1.4142z + 0.25}.$$

e. From Line 14 in Table 5.1, it follows that

$$X(z) = Z(e^{-0.1n}\cos(0.25\pi n)u(n)) = \frac{z(z - e^{-0.1}\cos(0.25\pi))}{z^2 - 2e^{-0.1}\cos(0.25\pi)z + e^{-0.2}}$$

$$= \frac{z(z - 0.6397)}{z^2 - 1.2794z + 0.8187}.$$

Example 5.4

Determine the z-transforms of the signal

$$x(n) = -\alpha^n u(-n-1) = \begin{cases} 0, & n \geq 0 \\ -\alpha^n, & n \leq -1 \end{cases}$$

Solution

$$X(z) = \sum_{n=-\infty}^{\infty} -\alpha^n u[-n-1]z^{-n} = -\sum_{n=-\infty}^{-1} \alpha^n z^{-n}$$

$$= -\sum_{n=1}^{\infty} \alpha^{-n} z^n = 1 - \sum_{n=0}^{\infty} (\alpha^{-1} z)^n.$$

For $|\alpha^{-1} z| < 1$ or $|z| < |\alpha|$, the series converges to

$$X(z) = 1 - \frac{1}{1 - \alpha^{-1} z} = \frac{1}{1 - \alpha z^{-1}} = \frac{z}{z - \alpha}, \quad |z| < |\alpha|.$$

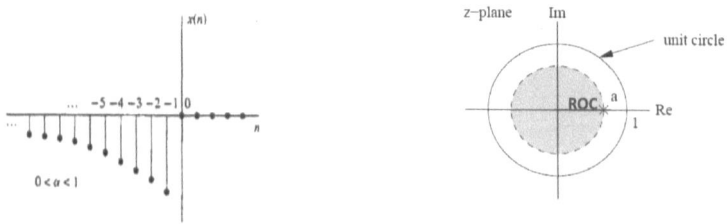

Figure 5.5 The one-sided function of Example 5.4 and its ROC

In signal processing, this definition is used when the signal is causal.

As analog filters are designed using the Laplace transform, recursive digital filters are developed with a parallel technique called the z-transform. The overall strategy of these two transforms is the same: probe the impulse response with sinusoids and exponentials to find the system's poles and zeros. The Laplace transforms deals with differential equations, the s-domain, and the s-plane. Correspondingly, the z-transform deals with difference equations, the z-domain, and the z-plane. However, the two techniques are not a mirror image of each other; the s-plane is arranged in a rectangular coordinate system, while the z-plane uses a polar format. Recursive digital filters are often designed by starting with one of the classic analog filters, such as the Butterworth, Chebyshev, or elliptic. A series of mathematical conversions are then used to obtain the desired digital filter. The Z transform of a discrete time system X[n] is defined as Power Series.

5.3 Bilateral Z-transform

The *bilateral* or *two-sided* Z-transform of a discrete-time signal x[n] is the function X(z) defined as:

$$X(z) = Z\{x[n]\} = \sum_{n=-\infty}^{\infty} x[n]z^{-n} \tag{5.6}$$

The two-sided (or bilateral) z-transform is again a complex function of a complex variable, meaning that it can take on complex values and that its argument is itself a complex variable. For the two-sided transform, we can consider again a few example sequences for which the sequence values are non-zero for both positive and negative index values.

Example 5.5
Consider the following sequence,

$$y[n] = a^n u[n] + b^n u[-n-1] = \begin{cases} a^n, & n \geq 0 \\ b^n, & n < 0. \end{cases}$$

Now, using the definition of the z-transform, we have for this sequence,

$$\begin{aligned}
Y(z) &= \sum_{n=-\infty}^{\infty} (a^n u[n] + b^n u[-n-1]) z^{-n} \\
&= \sum_{n=-\infty}^{\infty} (a^n u[n]) z^{-n} \sum_{n=-\infty}^{\infty} (b^n u[-n-1]) z^{-n} \\
&= \sum_{n=0}^{\infty} a^n z^{-n} \sum_{n=-\infty}^{-1} b^n z^{-n} \\
&= \sum_{n=0}^{\infty} \left(\frac{a}{z}\right)^n + \sum_{n=-\infty}^{-1} \left(\frac{b}{z}\right)^n \\
&= \frac{z}{z-a}, |z| > |a| + \sum_{n=-\infty}^{-1} \left(\frac{b}{z}\right)^n \\
&= \frac{z}{z-a}, |z| > |a| + \sum_{m=1}^{\infty} \left(\frac{z}{b}\right)^m \\
&= \frac{z}{z-a}, |z| > |a| + \frac{z}{b-z}, |z| < |b|,
\end{aligned}$$

where we must combine the two conditions on jzj, to ensure convergence of both summations in the expression. Otherwise, one of the terms in the expression will be invalid, and the resulting algebraic expression will not be meaningful. Hence, we have:

$$Y(z) = \frac{z}{z-a} + \frac{z}{b-z}, |a| < |z| < |b|.$$

5.4 Poles and Zeros in the Z-Plane

It is quite difficult to qualitatively analyse Z-transform, since mappings of their magnitude and phase or real part and imaginary part result in multiple mappings of 2-dimensional surfaces in 3-dimensional space. For this reason, it is very common to examine a plot of a transfer function's poles and zeros to try to gain a qualitative idea of what a system does.

Once the Z-transform of a system has been determined, one can use the information contained in function's polynomials to graphically represent the function and easily observe many defining characteristics. The Z-transform will have the below structure, based on Rational Functions:

$$H(z) = \frac{N(z)}{D(z)}$$

where $N(z)$ and $D(z)$ are polynomials in z. Assuming M^{th} order polynomials we have:

$$H(z) = \frac{b_0 + b_1 z + b_2 z^2 + \cdots + b_M z^M}{a_0 + a_1 z + a_2 z^2 + \cdots + a_M z^M} \tag{5.7}$$

In the complex domain a Mth order polynomial has exactly M zeros and we thus may write:

$$H(z) = K \frac{(z - z_1)(z - z_2) \cdots (z - z_M)}{(z - p_1)(z - p_2) \cdots (z - p_M)} \tag{5.8}$$

where the **zeros** z_i's are the zeros of $N(z)$ and the **poles** of the system, p_i's, are the zeros of $D(z)$.

Because a LTI system is completely characterized by its transfer function $H(z)$, the system is also completely characterized by its set of zeros and poles (together with a gain factor K). Plotting the zeros and poles in the complex plane gives the *Argand diagram* of the LTI system. In the Argand diagram we can also indicate the ROC of $H(z)$.

Therefore, the **Zeros** : are the value(s) for z where $N(z) = 0$. The complex frequencies that make the overall gain of the filter transfer function zero.

Poles : are the value(s) for z where $D(z) = 0$. The complex frequencies that make the overall gain of the filter transfer function infinite.

Example 5.6
Consider the LTI system with transfer function H(z)

$$H(z) = \frac{z^2 - 1.9z + 1}{z^2 - 1.8z + 0.9}$$

$$= \frac{(z - 0.95 - 0.31j)(z - 0.95 + 0.31j)}{(z - 0.9 - 0.3j)(z - 0.9 + 0.3j)}$$

The Argand diagram (plot of poles and zeros in the complex plane) and the frequency response $H(e^{j\omega})$ are sketched in figure 5.6.

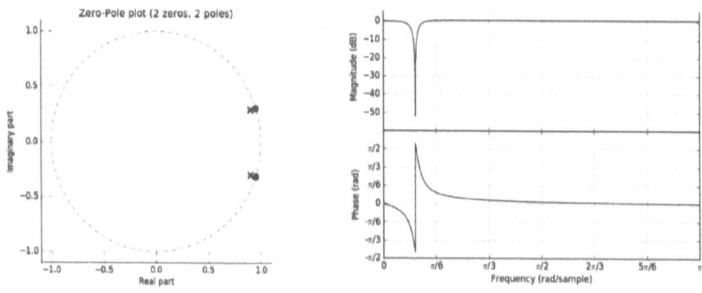

Figure 5.6 Poles and Zeroes in Z Plan

5.5 Properties of the z transform

For the following

$$Z\{f[n]\} = \sum_{n=0}^{n=\infty} f[n]z^{-n} = F(z) \qquad Z\{g_n\} = \sum_{n=0}^{n=\infty} g_n z^{-n} = G(z)$$

- **Linearity:**

$Z\{af_n + bg_n\} = aF(z) + bG(z)$. and ROC is $R_f \cap R_g$
which follows from definition of z-transform.

- **Time Shifting**

If we have $f[n] \Leftrightarrow F(z)$ then $f[n-n_0] \Leftrightarrow z^{-n_0} F(z)$
The ROC of $Y(z)$ is the same as $F(z)$ except that there are possible pole additions or deletions at $z = 0$ or $z = \infty$.

Proof:

Let $y[n] = f[n-n_0]$ then

$$Y(z) = \sum_{n=-\infty}^{\infty} f[n-n_0] z^{-n}$$

Assume $k = n - n0$ then $n = k + n0$, substituting in the above equation we have:

$$Y(z) = \sum_{k=-\infty}^{\infty} f[k] z^{-k-n_0} = z^{-n_0} F[z]$$

- **Multiplication by an Exponential Sequence**

Let $y[n] = z_0^n f[n]$ then $Y(z) = X\left(\dfrac{z}{z_0}\right)$

The consequence is pole and zero locations are scaled by z_0. If the ROC of $FX(z)$ is $r_R < |z| < r_L$, then the ROC of $Y(z)$ is

$$r_R < |z/z_0| < r_L, \; i.e., \; |z_0|r_R < |z| < |z_0|r_L$$

Proof:

$$Y(z) = \sum_{n=-\infty}^{\infty} z_0^n x[n] z^{-n} = \sum_{n=-\infty}^{\infty} x[n] \left(\frac{z}{z_0}\right)^{-n} = X\left(\frac{z}{z_0}\right)$$

The consequence is pole and zero locations are scaled by z_0. If the ROC of $X(z)$ is $rR < |z| < rL$, then the ROC of $Y(z)$ is

$$r_R < |z/z_0| < r_L, \; i.e., \; |z_0|r_R < |z| < |z_0|r_L$$

- **Differentiation of $X(z)$**

If we have $f[n] \Leftrightarrow F(z)$ then $nf[n] \xleftarrow{z} -z\dfrac{dF(z)}{z}$ and ROC = R_f

Proof:

$$F(z) = \sum_{n=-\infty}^{\infty} f[n] z^{-n}$$

$$-z\frac{dF(z)}{dz} = -z \sum_{n=-\infty}^{\infty} -n f[n] z^{-n-1} = \sum_{n=-\infty}^{\infty} -n f[n] z^{-n}$$

$$-z\frac{dF(z)}{dz} \xleftarrow{z} nf[n]$$

- **Conjugation of a Complex Sequence**

If we have $f[n] \Leftrightarrow F(z)$ then $f^*[n] \xleftarrow{z} F^*(z^*)$ and ROC = R_f

Proof:

Let $y[n] = f^*[n]$, then

$$Y(z) = \sum_{n=-\infty}^{\infty} f^*[n]z^{-n} = \left(\sum_{n=-\infty}^{\infty} f[n][z^*]^{-n}\right)^* = F^*(z^*)$$

- **Time Reversal**

If we have $f[n] \Leftrightarrow F(z)$ then $f^*[-n] \xleftarrow{z} F^*(1/z^*)$

Let $y[n] = f^*[-n]$, then

$$Y(z) = \sum_{n=-\infty}^{\infty} f^*[-n]z^{-n} = \left(\sum_{n=-\infty}^{\infty} f[-n][z^*]^{-n}\right)^* = \left(\sum_{k=-\infty}^{\infty} f[k](1/z^*)^{-k}\right)^* = F^*(1/z^*)$$

If the ROC of $F(z)$ is $r_R < |z| < r_L$, then the ROC of $Y(z)$ is

$$r_R < |1/z^*| < r_L \quad \text{i.e.,} \quad \frac{1}{r_R} > |z| > \frac{1}{r_L}$$

When the time reversal is without conjugation, it is easy to show

$$f[-n] \xleftarrow{z} F(1/z) \quad \text{and ROC is } \frac{1}{r_R} > |z| > \frac{1}{r_L}$$

A comprehensive summery for the z-transform properties is shown in Table 2

The following properties of z-transforms listed in Table 5.2 are well known in the field of digital signal analysis. The reader will be asked to prove some of these properties in the exercises.

Table 5.2 Summery of z-transform properties

		Sequence	z - transform
	definition	$x_n = x[n]$	$X(z) = \sum_{n=0}^{\infty} x_n z^{-n}$
1	addition	$x_n + y_n$	$X(z) + Y(z)$
2	constant multiple	$c\, x_n$	$c\, X(z)$
3	linearity	$c\, x_n + d\, y_n$	$c\, X(z) + d\, Y(z)$
4	delayed unit step	$u[n-m]$	$\dfrac{z^{1-m}}{z-1}$
5	time delay 1 tap	$x_{n-1} u[n-1]$	$\dfrac{1}{z} X(z)$
6	time delayed shift	$x_{n-m} u[n-m]$	$z^{-m} X(z)$
7	forward 1 tap	x_{n+1}	$z\,(X(z) - x_0)$
8	forward 2 taps	x_{n+2}	$z^2\,(X(z) - x_0 - x_1 z^{-1})$
9	time forward	x_{n+m}	$z^m \left(X(z) - \sum_{i=0}^{m-1} x_i z^{-i} \right)$
10	complex translation	$e^{an} x_n$	$X(z\, e^{-a})$
11	frequency scale	$b^n x_n$	$X\left(\dfrac{z}{b}\right)$
12	differentiation	$n\, x_n$	$-z X'(z)$
13	integration	$\dfrac{1}{n} x_n$	$-\int \dfrac{X(z)}{z} dz$
14	integration shift	$\dfrac{1}{n+m} x_n$	$-z^{-m} \int \dfrac{X(z)}{z^{m+1}} dz$
15	discrete time convolution	$x_n * y_n = \sum_{i=0}^{n} x_i y_{n-i}$	$X(z)\, Y(z)$
16	convolution with $y_n = 1$	$\sum_{i=0}^{n} x_i$	$\dfrac{z}{z-1} X(z)$
17	initial time	x_0	$\lim_{z \to \infty} X(z)$
18	final value	$\lim_{n \to \infty} x_n$	$\lim_{z \to 1} (z-1) X(z)$

5.6 Region of Convergence for the Z-Transform

The region of convergence, known as the **ROC**, is important to understand because it defines the region where the z-transform exists. The **z-transform** of a sequence is defined as:

$$X(z) = \sum_{n=-\infty}^{\infty} x[n]z^{-n} \qquad (5.9)$$

The ROC for a given $x[n]$, is defined as the range of z for which the z-transform converges. Since the z-transform is a **power series**, it converges when $x[n]z^{-n}$ is absolutely summable. Stated differently,

$$\sum_{n=-\infty}^{\infty} |x[n]z^{-n}| < \infty \qquad (5.10)$$

must be satisfied for convergence.

5.6.1 Properties of the Region of Convergence

The Region of Convergence has a number of properties that are dependent on the characteristics of the signal, $x[n]$

- **The ROC cannot contain any poles.** By definition a pole is a where $x(z)$ is infinite. Since $X(z)$ must be finite for all zz for convergence, there cannot be a pole in the ROC.
- **If $x[n]$ is a finite-duration sequence, then the ROC is the entire z-plane, except possibly $z = 0$ or $|z| = \infty|$.** A **finite-duration sequence** is a sequence that is nonzero in a finite interval $n1 \leq n \leq n2$. As long as each value of $x[n]$ is finite then the sequence will be absolutely summable. When $n2 > 0$ there will be a $z - 1$ term and thus the ROC will not include $z = 0$. When n1<0n1<0 then the sum will

be infinite and thus the ROC will not include $|z| = \infty$. On the other hand, when $n2 \leq 0$ then the ROC will include $z=0$ and when $n1 \geq 0$ the ROC will include $|z| = \infty$. With these constraints, the only signal, then, whose ROC is the entire z-plane is $x[n] = c\delta[n]$.

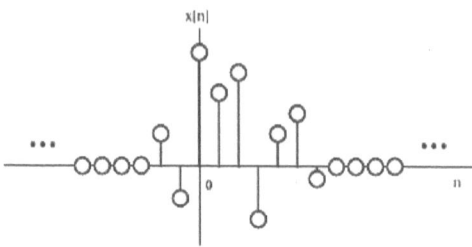

Figure 5.7 An example of a finite duration sequence.

The next properties apply to infinite duration sequences. As noted above, the z-transform converges when $|X(z)| < \infty$ So we can write

$$|X(z)| = \left|\sum_{n=-\infty}^{\infty} x[n]z^{-n}\right| \leq \sum_{n=-\infty}^{\infty} |x[n]z^{-n}| = \sum_{n=-\infty}^{\infty} |x[n]|(|z|)^{-n}$$

We can then split the infinite sum into positive-time and negative-time portions. So

$$|X(z)| \leq N(z) + P(z)$$

where

$$N(z) = \sum_{n=-\infty}^{-1} |x[n]|(|z|)^{-n}$$

And

$$P(z) = \sum_{n=0}^{\infty} |x[n]|(|z|)^{-n}$$

In order for $|X(z)|$ to be finite, $|x[n]||x[n]|$ must be bounded. Let us then set

$$|x(n)| \leq C_1 r_1^n$$

for

$$n < 0$$

and

$$|x(n)| \leq C_2 r_2^n$$

For

$$n \geq 0$$

From this some further properties can be derived:

- **If $x[n]$ is a right-sided sequence, then the ROC extends outward from the outermost pole in $X(z)$.** A **right-sided sequence** is a sequence where $x[n] = 0$ for $n < n1 < \infty$. Looking at the positive-time portion from the above derivation, it follows that

$$P(z) \leq C_2 \sum_{n=0}^{\infty} r_2^n (|z|)^{-n} = C_2 \sum_{n=0}^{\infty} \left(\frac{r_2}{|z|}\right)^n$$

Thus in order for this sum to converge, $|z| > r2$, and therefore the ROC of a right-sided sequence is of the form $|z| > r2.$.

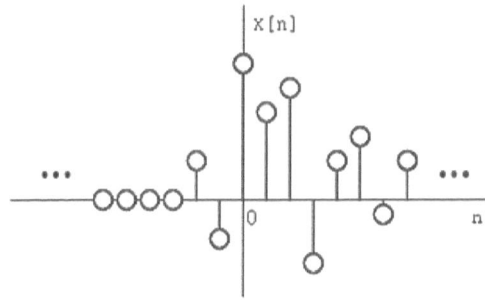

Figure 5.8 A right-sided sequence.

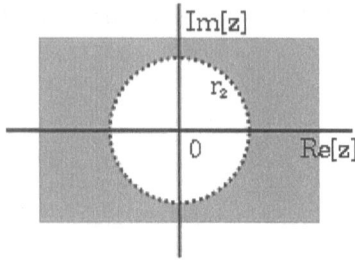

Figure 5.9 The ROC of a right-sided sequence.

- **If $x[n]$ is a left-sided sequence, then the ROC extends inward from the innermost pole in $X(z)$. A left-sided sequence is a sequence where $x[n] = 0$ for $n > n2 > -\infty$.** Looking at the negative-time portion from the above derivation, it follows that

$$N(z) \leq C_1 \sum_{n=-\infty}^{-1} r_1^n(|z|)^{-n} = C_1 \sum_{n=-\infty}^{-1} \left(\frac{r_1}{|z|}\right)^n = C_1 \sum_{k=1}^{\infty} \left(\frac{|z|}{r_1}\right)^k$$

Thus in order for this sum to converge, $|z| < r1$, and therefore the ROC of a left-sided sequence is of the form $|z| < r1$.

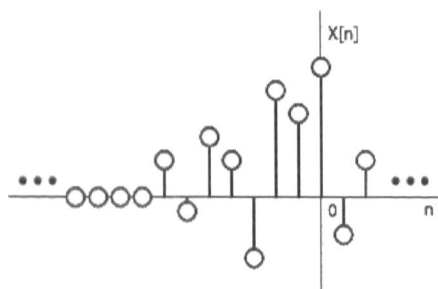

Figure 5.10 A left-sided sequence.

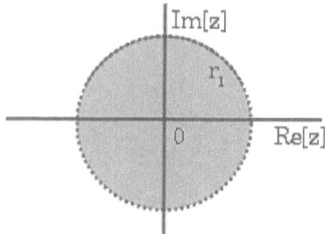

Figure 5.11 The ROC of a left-sided sequence.

- **If $x[n]$ is a two-sided sequence, the ROC will be a ring in the z-plane that is bounded on the interior and exterior by a pole.** A two-sided sequence is a sequence with infinite duration in the positive and negative directions. From the derivation of the above two properties, it follows that if $-r2 < |z| < r2$ converges, then both the positive-time and negative-time portions converge and thus $X(z)$ converges as well. Therefore the ROC of a two-sided sequence is of the form $-r2 < |z| < r2$.

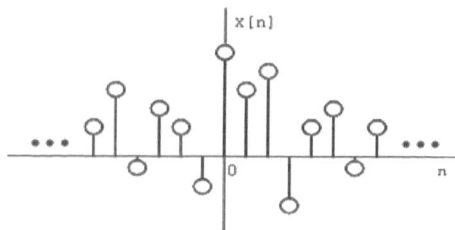

Figure 5.12 A two-sided sequence.

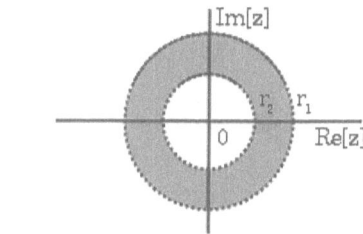

Figure 5.13 The ROC of a two-sided sequence.

Example 5.7.

Consider the function.

$$x_1[n] = \left(\frac{1}{2}\right)^n u[n] + \left(\frac{1}{4}\right)^n u[n]$$

Determine the ROC.

Solution

The z-transform of $\left(\frac{1}{2}\right)^n u[n]$ is $\frac{z}{z-\frac{1}{2}}$ with an ROC at $|z| > \frac{1}{2}$,

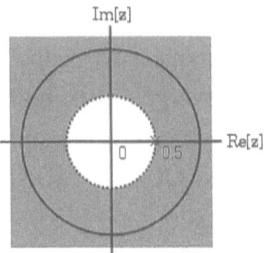

Figure 5.14 The ROC of $\left(\frac{1}{2}\right)^n u[n]$

The z-transform of $\left(\frac{-1}{4}\right)^n u[n]$ is $\frac{z}{z-\frac{-1}{4}}$ with an ROC at $|z| > \frac{-1}{4}$,

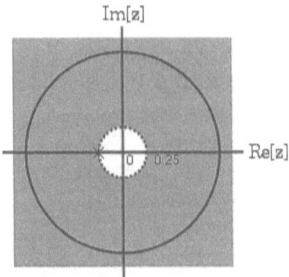

Figure 5.15 The ROC of $\left(\frac{-1}{4}\right)^n u[n]$

Due to linearity,

$$X_1[z] = \frac{z}{z-\frac{1}{2}} + \frac{z}{z+\frac{1}{4}}$$

$$= \frac{2z\left(z-\frac{1}{8}\right)}{\left(z-\frac{1}{2}\right)\left(z+\frac{1}{4}\right)}$$

By observation *there* are two zeros, at $z = 0$ and $z = 18$, and two poles, at $z = 12$, and $z = -14$. Following the *above* properties, the ROC is $|z| > 12$.

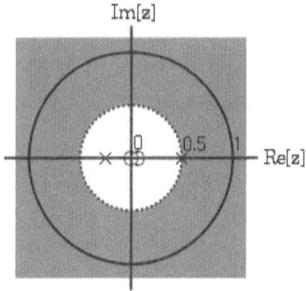

Figure 5.15 The ROC of $(\frac{1}{2})^n u[n] + (\frac{-1}{4})^n u[n]$

5.7 Inverse z-Transform

The z-transform of the sequence $x(n)$ and the inverse z-transform of the function $X(z)$ are defined as, respectively,

$$X(z) = Z\{x[n]\} \qquad (5.11)$$

and $x[n] = Z^{-1}\{X(z)\}$ (5.12)

where $Z(\ )$ is the z-transform operator, while $Z^{-1}(\ )$ is the inverse z-transform operator. The inverse z-transform may be obtained by at least three methods:

1. Power series expansion

2. Partial fraction expansion and look-up table.
3. Contour integration.

5.7.1 Power Series Method

The z-transform is a power series expansion,

$$X(z) = \sum_{n=-\infty}^{\infty} x(n)z^{-n} = \cdots + x(-2)z^2 + x(-1)z + x(0) + x(1)z^{-1} + x(2)z^{-2} + \cdots$$

where the sequence values $x(n)$ are the coefficients of z^{-n} in the expansion. Therefore, if we can find the power series expansion for $X(z)$, the sequence values $x(n)$ may be found by simply picking off the coefficients of z^{-n}.

Also, it called long division expansion, consider the function

$$X(z) = \frac{1}{1 - az^{-1}}, \qquad |z| > |a|.$$

Since the ROC is the exterior of a circle, the sequence is right-sided. We Therefore divide to get a power series in powers of z^{-1}:

$$\begin{array}{r} 1+az^{-1}+a^2z^{-2}+\cdots \\ 1-az^{-1}\overline{)\,1\phantom{-az^{-1}-a^2z^{-2}}} \\ \underline{1-az^{-1}} \\ az^{-1} \\ \underline{az^{-1}-a^2z^{-2}} \\ a^2z^{-2}+\cdots \end{array}$$

Or

$$\frac{1}{1 - az^{-1}} = 1 + az^{-1} + a^2z^{-2} + \cdots.$$

Therefore

$$x[n] = a^n u[n].$$

Example 5.8
Consider the transform.

$$X(z) = \log(1 + az^{-1}), \qquad |z| > |a|.$$

Using the power series expansion for log $(1 + x)$, with $|x| < 1$, gives

$$X(z) = \sum_{n=1}^{\infty} \frac{(-1)^{n+1} a^n z^{-n}}{n}.$$

The corresponding sequence is therefore;

$$x[n] = \begin{cases} (-1)^{n+1} \frac{a^n}{n} & n \geq 1 \\ 0 & n \leq 0. \end{cases}$$

5.7.2 Partial Fraction Expansion

Now, we are ready to deal with the inverse z-transform using the partial fraction expansion and look-up table. The general procedure is as follows:

1. Eliminate the negative powers of z for the z-transform function $X(z)$.
2. Determine the rational function $X(z)/z$ (assuming it is proper) and apply the partial fraction expansion to the determined rational function $X(z)/z$ using the formula in Table 5.3.

3. Multiply the expanded function $X(z)/z$ by z on both sides of the equation to obtain $X(z)$.
4. Apply the inverse z-transform using Table 5.1.

The partial fraction format and the formula for calculating the constants are listed in Table 5.3.:

Table 5.3 Partial fraction(s) and formulas for constant(s).

Partial fraction with the first-order real pole:
$$\frac{R}{z-p} \qquad R = (z-p)\frac{X(z)}{z}\Big|_{z=p}$$

Partial fraction with the first-order complex poles:
$$\frac{Az}{(z-P)} + \frac{A^*z}{(z-P^*)} \qquad A = (z-P)\frac{X(z)}{z}\Big|_{z=P}$$

P^* = complex conjugate of P
A^* = complex conjugate of A

Partial fraction with mth-order real poles:
$$\frac{R_m}{(z-p)} + \frac{R_{m-1}}{(z-p)^2} + \cdots + \frac{R_1}{(z-p)^m} \qquad R_k = \frac{1}{(k-1)!}\frac{d^{k-1}}{dz^{k-1}}\left((z-p)^m\frac{X(z)}{z}\right)\Big|_{z=p}$$

Let us considers the situation of the z-transform function having first order poles.

Example 5.9.

a. Find the inverse of the following z-transform:

$$X(z) = \frac{1}{(1-z^{-1})(1-0.5z^{-1})}.$$

Solution:

Eliminating the negative power of z by multiplying the numerator and denominator by z^2 yields

$$X(z) = \frac{z^2}{z^2(1-z^{-1})(1-0.5z^{-1})}.$$

$$= \frac{z^2}{(z-1)(z-0.5)}$$

Dividing both sides by z leads to

$$\frac{X(z)}{z} = \frac{z}{(z-1)(z-0.5)}.$$

Again, we write

$$\frac{X(z)}{z} = \frac{A}{(z-1)} + \frac{B}{(z-0.5)}.$$

Then A and B are constants found using the formula in Table 5.3, that is,

$$A = (z-1)\frac{X(z)}{z}\bigg|_{z=1} = \frac{z}{(z-0.5)}\bigg|_{z=1} = 2,$$

$$B = (z-0.5)\frac{X(z)}{z}\bigg|_{z=0.5} = \frac{z}{(z-1)}\bigg|_{z=0.5} = -1.$$

Thus

$$\frac{X(z)}{z} = \frac{2}{(z-1)} + \frac{-1}{(z-0.5)}.$$

Multiplying z on both sides gives

$$X(z) = \frac{2z}{(z-1)} + \frac{-z}{(z-0.5)}.$$

Using Table 5.1 of the z-transform pairs, the resultant x(n) is :

$$x(n) = 2u(n) - (0.5)^n u(n).$$

Let us consider the case where $X(z)$ has first-order complex poles.

Example 5.10.
Find $y(n)$ if

$$Y(z) = \frac{z^2(z+1)}{(z-1)(z^2-z+0.5)}.$$

Solution
Dividing $Y(z)$ by z, we have

$$\frac{Y(z)}{z} = \frac{z(z+1)}{(z-1)(z^2-z+0.5)}.$$

Applying the partial fraction expansion leads to

$$\frac{Y(z)}{z} = \frac{B}{z-1} + \frac{A}{(z-0.5-j0.5)} + \frac{A^*}{(z-0.5+j0.5)}.$$

We first find B:

$$B = (z-1)\frac{Y(z)}{z}\bigg|_{z=1} = \frac{z(z+1)}{(z^2-z+0.5)}\bigg|_{z=1} = \frac{1\times(1+1)}{(1^2-1+0.5)} = 4$$

Notice that A and A^* form a complex conjugate pair. We determine A as follows:

$$A = (z-0.5-j0.5)\frac{Y(z)}{z}\bigg|_{z=0.5+j0.5} = \frac{z(z+1)}{(z-1)(z-0.5+j0.5)}\bigg|_{z=0.5+j0.5}$$

$$A = \frac{(0.5+j0.5)(0.5+j0.5+1)}{(0.5+j0.5-1)(0.5+j0.5-0.5+j0.5)} = \frac{(0.5+j0.5)(1.5+j0.5)}{(-0.5+j0.5)j}$$

Using the polar form, we get

$$A = \frac{(0.707\angle 45°)(1.58114\angle 18.43°)}{(0.707\angle 135°)(1\angle 90°)} = 1.58114\angle -161.57°$$

$$A^* = \bar{A} = 1.58114\angle 161.57°.$$

Assume that a first-order complex pole has the form

$$P = 0.5 + 0.5j = |P|\angle\theta = 0.707\angle 45° \text{ and } P^* = |P|\angle -\theta = 0.707\angle -45°.$$

We have

$$Y(z) = \frac{4z}{z-1} + \frac{Az}{(z-P)} + \frac{A^*z}{(z-P^*)}.$$

Applying the inverse z-transform from Table 5.1 leads to

$$y(n) = 4Z^{-1}\left(\frac{z}{z-1}\right) + Z^{-1}\left(\frac{Az}{(z-P)} + \frac{A^*z}{(z-P^*)}\right).$$

Using the previous formula, the inversion and subsequent simplification yield

$$y(n) = 4u(n) + 2|A|(|P|)^n \cos(n\theta + \phi)u(n)$$
$$= 4u(n) + 3.1623(0.7071)^n \cos(45°n - 161.57°)u(n)$$

Now, let us deal with the real repeated poles is the next example.

Example 5.11
Find x(n) if

$$X(z) = \frac{z^2}{(z-1)(z-0.5)^2}.$$

Solution

Dividing both sides of the previous z-transform by z yields

$$\frac{X(z)}{z} = \frac{z}{(z-1)(z-0.5)^2} = \frac{A}{z-1} + \frac{B}{z-0.5} + \frac{C}{(z-0.5)^2},$$

where $A = (z-1)\frac{X(z)}{z}\bigg|_{z=1} = \frac{z}{(z-0.5)^2}\bigg|_{z=1} = 4.$

Using the formulas for mth-order real poles in Table 5.3, where m = 2 and p = 0:5, to determine B and C yields

$$B = R_2 = \frac{1}{(2-1)!} \frac{d}{dz}\left\{(z-0.5)^2 \frac{X(z)}{z}\right\}_{z=0.5}$$

$$= \frac{d}{dz}\left(\frac{z}{z-1}\right)\bigg|_{z=0.5} = \frac{-1}{(z-1)^2}\bigg|_{z=0.5} = -4$$

$$C = R_1 = \frac{1}{(1-1)!} \frac{d^0}{dz^0}\left\{(z-0.5)^2 \frac{X(z)}{z}\right\}_{z=0.5}$$

$$= \frac{z}{z-1}\bigg|_{z=0.5} = -1.$$

Then $X(z) = \frac{4z}{z-1} + \frac{-4z}{z-0.5} + \frac{-1z}{(z-0.5)^2}.$

The inverse z-transform for each term on the right-hand side of the above Equation can be achieved by the result listed in Table 5.1, that is,

$$Z^{-1}\left\{\frac{z}{z-1}\right\} = u(n),$$

$$Z^{-1}\left\{\frac{z}{z-0.5}\right\} = (0.5)^n u(n),$$

$$Z^{-1}\left\{\frac{z}{(z-0.5)^2}\right\} = 2n(0.5)^n u(n).$$

From these results, it follows that

$$x(n) = 4u(n) - 4(0.5)^n u(n) - 2n(0.5)^n u(n).$$

5.7.3 Contour integration.

Another approach that may be used to find the inverse z-transform of $X(z)$ is to use contour integration. This procedure relies on Cauchy's integral theorem, which states that if C is a closed contour that encircles the origin in a counterclockwise direction,

$$\frac{1}{2\pi j}\oint_C z^{-k}dz = \begin{cases} 1 & k=1 \\ 0 & k \neq 1 \end{cases} \quad (5.13)$$

With

$$X(z) = \sum_{n=-\infty}^{\infty} x(n) z^{-n}$$

Cauchy's integral theorem may be used to show that the coefficients $x(n)$ may be found from $X(z)$ as follows:

$$x(n) = \frac{1}{2\pi j} \oint_C X(z) z^{n-1} dz$$

where C is a closed contour within the region of convergence of $X(z)$ that encircles the origin in a counterclockwise direction. Contour integrals of this form may often by evaluated with the help of Cauchy's residue theorem,

$$x(n) = \frac{1}{2\pi j} \oint_C X(z) z^{n-1} dz = \sum [\text{residues of } X(z) z^{n-1} \text{ at the poles inside } C]$$

If $X(z)$ is a rational function of z with a first-order pole at $z = \alpha_k$,

$$\text{Res}\left[X(z) z^{n-1} \text{ at } z = \alpha_k\right] = \left[(1 - \alpha_k z^{-1}) X(z) z^{n-1}\right]_{z=\alpha_k}$$

Contour integration is particularly useful if only a few values of $x(n)$ are needed.

5.8 Transfer Function in the Z-domain

A LTI system is completely characterized by its impulse response $h[n]$ or equivalently the Z-transform of the impulse response $H(z)H(z)$ which is called the transfer function. Remember:

$$x[n] * h[n] \xrightarrow{Z} X(z) H(z).$$

In case the impulse response is given to define the LTI system we can simply calculate the Z-transform to obtain :math:` H(x).

In case the system is defined with a difference equation we could first calculate the impulse response and then calculating the Z-transform. But it is far easier to calculate the Z-transform of both sides of the difference equation.

As an example consider the following difference equation:

$$y[n] = 1.5y[n-1] - 0.5y[n-2] + 0.5x[n].$$

The Z-transform is a linear transform we can apply the Z-transform to both sides of the above equation and obtain:

$$Y(z) = 1.5z^{-1}Y(z) - 0.5z^{-2}Y(z) + 0.5X(z)$$

This can be rewritten as:

$$H(z) = \frac{Y(z)}{X(z)} = \frac{0.5}{1 - 1.5z^{-1} + 0.5z^{-2}} = \frac{z^2}{2z^2 - 3z + 1}$$

5.9 Application to signal processing

5.9.1 Solution of Difference Equations Using the z-Transform

To solve a difference equation with initial conditions, we have to deal with time shifted sequences such $y(n-1), y(n-2), \ldots, y(n-m)$, and so on. Let us examine the z-transform of these terms. Using the definition of the z-transform, we have

$$Z(y(n-1)) = \sum_{n=0}^{\infty} y(n-1)z^{-n}$$
$$= y(-1) + y(0)z^{-1} + y(1)z^{-2} + \ldots$$
$$= y(-1) + z^{-1}(y(0) + y(1)z^{-1} + y(2)z^{-2} + \ldots)$$

It holds that

$$Z(y(n-1)) = y(-1) + z^{-1}Y(z).$$

Similarly, we can have

$$Z(y(n-2)) = \sum_{n=0}^{\infty} y(n-2)z^{-n}$$
$$= y(-2) + y(-1)z^{-1} + y(0)z^{-2} + y(1)z^{-3} + \ldots$$
$$= y(-2) + y(-1)z^{-1} + z^{-2}(y(0) + y(1)z^{-1} + y(2)z^{-2} + \ldots)$$
$$Z(y(n-2)) = y(-2) + y(-1)z^{-1} + z^{-2}Y(z)$$
$$Z(y(n-m)) = y(-m) + y(-m+1)z^{-1} + \ldots + y(-1)z^{-(m-1)}$$
$$+ z^{-m}Y(z),$$

where $y(-m), y(-m+1), \ldots, y(-1)$ are the initial conditions. If all initial conditions are considered to be zero, that is,

$$y(-m) = y(-m+1) = \ldots y(-1) = 0, \qquad (5:14)$$

then Equation (5.12) becomes

$$Z(y(n-m)) = z^{-m}Y(z),$$

The following two examples serve as illustrations of applying the z-transform to find the solutions of the difference equations. The procedure is:

1. Apply z-transform to the difference equation.
2. Substitute the initial conditions.
3. Solve for the difference equation in z-transform domain.
4. Find the solution in time domain by applying the inverse z-transform.

Example 5.12.
A digital signal processing (DSP) system is described by the difference equation

$$y(n) - 0.5y(n-1) = 5(0.2)^n u(n).$$

Determine the solution when the initial condition is given by $y(-1) = 1$.

Solution
Applying the z-transform on both sides of the difference equation, we have

$$Y(z) - 0.5\left(y(-1) + z^{-1}Y(z)\right) = 5Z(0.2^n u(n)).$$

Substituting the initial condition and

$$Z(0.2^n u(n)) = z/(z - 0.2),$$

we achieve

$$Y(z) - 0.5\left(1 + z^{-1}Y(z)\right) = 5z/(z - 0.2).$$

Simplification yields

$$Y(z) - 0.5z^{-1}Y(z) = 0.5 + 5z/(z - 0.2).$$

Factoring out $Y(z)$ and combining the right-hand side of the equation, it follows that

$$Y(z)(1 - 0.5z^{-1}) = (5.5z - 0.1)/(z - 0.2).$$

Then we obtain

$$Y(z) = \frac{(5.5z - 0.1)}{(1 - 0.5z^{-1})(z - 0.2)} = \frac{z(5.5z - 0.1)}{(z - 0.5)(z - 0.2)}.$$

Using the partial fraction expansion method leads to

$$\frac{Y(z)}{z} = \frac{5.5z - 0.1}{(z - 0.5)(z - 0.2)} = \frac{A}{z - 0.5} + \frac{B}{z - 0.2},$$

Where

$$A = (z - 0.5)\frac{Y(z)}{z}\bigg|_{z=0.5} = \frac{5.5z - 0.1}{z - 0.2}\bigg|_{z=0.5} = \frac{5.5 \times 0.5 - 0.1}{0.5 - 0.2} = 8.8333,$$

$$B = (z - 0.2)\frac{Y(z)}{z}\bigg|_{z=0.2} = \frac{5.5z - 0.1}{z - 0.5}\bigg|_{z=0.2} = \frac{5.5 \times 0.2 - 0.1}{0.2 - 0.5} = -3.3333.$$

Thus

$$Y(z) = \frac{8.8333z}{(z - 0.5)} + \frac{-3.3333z}{(z - 0.2)},$$

which gives the solution as

$$y(n) = 8.3333(0.5)^n u(n) - 3.3333(0.2)^n u(n).$$

Example 5.13.

A relaxed (zero initial conditions) DSP system is described by the difference Equation

$$y(n) + 0.1y(n-1) - 0.2y(n-2) = x(n) + x(n-1).$$

a. Determine the impulse response $y(n)$ due to the impulse sequence $x(n) = \delta(n)$
b. Determine system response $y(n)$ due to the unit step function excitation,
where $u(n) = 1$ for $n >= 0$.

Solution:

a. Applying the z-transform on both sides of the difference equations we yield

$$Y(z) + 0.1\,Y(z)z^{-1} - 0.2\,Y(z)z^{-2} = X(z) + X(z)z^{-1}.$$

Factoring out Y(z) on the left side and substituting X(z) = Z(δ(n)) = 1 to the right side we achieve:

$$Y(z)(1 + 0.1z^{-1} - 0.2z^{-2}) = 1(1 + z^{-1}).$$

Then Y(z) can be expressed as

$$Y(z) = \frac{1 + z^{-1}}{1 + 0.1z^{-1} - 0.2z^{-2}}.$$

To obtain the impulse response, which is the inverse z-transform of the transfer function, we multiply the numerator and denominator by $Z^{\wedge}2$

Thus

$$Y(z) = \frac{z^2 + z}{z^2 + 0.1z - 0.2} = \frac{z(z+1)}{(z-0.4)(z+0.5)}.$$

Using the partial fraction expansion method leads to

$$\frac{Y(z)}{z} = \frac{z+1}{(z-0.4)(z+0.5)} = \frac{A}{z-0.4} + \frac{B}{z+0.5},$$

where $A = (z-0.4)\dfrac{Y(z)}{z}\bigg|_{z=0.4} = \dfrac{z+1}{z+0.5}\bigg|_{z=0.4} = \dfrac{0.4+1}{0.4+0.5} = 1.5556$

$B = (z+0.5)\dfrac{Y(z)}{z}\bigg|_{z=-0.5} = \dfrac{z+1}{z-0.4}\bigg|_{z=-0.5} = \dfrac{-0.5+1}{-0.5-0.4} = -0.5556.$

Thus

$$Y(z) = \frac{1.5556z}{(z-0.4)} + \frac{-0.5556z}{(z+0.5)},$$

which gives the impulse response:

$$y(n) = 1.5556(0.4)^n u(n) - 0.5556(-0.5)^n u(n).$$

b. To obtain the response due to a unit step function, the input sequence is set to be $x(n) = u(n)$

and the corresponding z-transform is given by

$$X(z) = \frac{z}{z-1},$$

and notice that

$$Y(z) + 0.1Y(z)z^{-1} - 0.2Y(z)z^{-2} = X(z) + X(z)z^{-1}.$$

Then the z-transform of the output sequence y(n) can be yielded as

$$Y(z) = \left(\frac{z}{z-1}\right)\left(\frac{1+z^{-1}}{1+0.1z^{-1}-0.2z^{-2}}\right) = \frac{z^2(z+1)}{(z-1)(z-0.4)(z+0.5)}.$$

Using the partial fraction expansion method as before gives

$$Y(z) = \frac{2.2222z}{z-1} + \frac{-1.0370z}{z-0.4} + \frac{-0.1852z}{z+0.5},$$

and the system response is found by using Table 5.1:

$$y(n) = 2.2222u(n) - 1.0370(0.4)^n u(n) - 0.1852(-0.5)^n u(n)$$

5.9.2 Analysis of Linear Discrete Systems

We are able to obtain the transfer function by ignoring initial conditions. The result is

$$H(z) = \frac{Y(z)}{F(z)} = \frac{\sum_{k=0}^{L} b_k z^{-k}}{\sum_{k=0}^{N} a_k z^{-k}} = \text{transfer function}$$

where $H(z)$ is the transform of the impulse response of a discrete system.

Stability
Using the convolution relation between input and output of a discrete systems, we obtain

$$|y(n)| = \left|\sum_{k=0}^{n} h(k) f(n-k)\right| \le M \sum_{k=0}^{\infty} |h(k)| < \infty$$

where M is the maximum value of $f(n)$. The above inequality specifies that a discrete system is stable if to a finite input the absolute sum of its impulse response is finite. From the properties of the Z-transform, the ROC of the impulse response is $|z| > 1$. Hence, all the poles of $H(z)$ of a stable system lie inside the unit circle.

Causality
A system is causal if $h(n) - 0$ for $n < 0$. From the properties of the Z-transform, $H(z)$ is regular in the ROC and at the infinity point. For rational functions the numerator polynomial has to be at most of the same degree as the polynomial in the denominator.

The Paley-Wiener theorem provides the necessary and sufficient conditions that a frequency response characteristic $H(\omega)$ must satisfy in order for the resulting filter to be causal.

Problems

5.1. Find the z-transform for each of the following sequences:

a. $x(n) = 4u(n)$

b. $x(n) = (-0.7)^n u(n)$

c. $x(n) = 4e^{-2n} u(n)$

d. $x(n) = 4(0.8)^n \cos(0.1\pi n) u(n)$

e. $x(n) = 4e^{-3n} \sin(0.1\pi n) u(n)$.

5.2. Using the properties of the z-transform, find the z-transform for each of the following sequences:

a. $x(n) = u(n) + (0.5)^n u(n)$

b. $x(n) = e^{-3(n-4)} \cos(0.1\pi(n-4)) u(n-4)$, where $u(n-4) = 1$ for $n \geq 4$ while $u(n-4) = 0$ for $n < 4$.

5.3 Given two sequences,

$$x_1(n) = 5\delta(n) - 2\delta(n-2) \text{ and}$$
$$x_2(n) = 3\delta(n-3),$$

determine the z-transform of convolution of the two sequences using the convolution property of z-transform

$$X(z) = X_1(z)X_2(z);$$

determine convolution by the inverse z-transform from the result in (a)

$$x(n) = Z^{-1}(X_1(z)X_2(z)).$$

5.4. Using Table 5.1 and z-transform properties, find the inverse z-transform for each of the following functions:

a. $X(z) = 4 - \dfrac{10z}{z-1} - \dfrac{z}{z+0.5}$

b. $X(z) = \dfrac{-5z}{(z-1)} + \dfrac{10z}{(z-1)^2} + \dfrac{2z}{(z-0.8)^2}$

c. $X(z) = \dfrac{z}{z^2 + 1.2z + 1}$

d. $X(z) = \dfrac{4z^{-4}}{z-1} + \dfrac{z^{-1}}{(z-1)^2} + z^{-8} + \dfrac{z^{-5}}{z-0.5}$

5.5. Using the partial fraction expansion method, find the inverse of the following z-transforms:

a. $X(z) = \dfrac{1}{z^2 - 0.3z - 0.04}$

b. $X(z) = \dfrac{z}{(z-0.2)(z+0.4)}$

c. $X(z) = \dfrac{z}{(z+0.2)(z^2 - z + 0.5)}$

d. $X(z) = \dfrac{z(z+0.5)}{(z-0.1)^2(z-0.6)}$

5.6. A system is described by the difference equation

$$y(n) + 0.5y(n-1) = 2(0.8)^n u(n).$$

Determine the solution when the initial condition is y(-1) = 2.

5.7. A system is described by the difference equation

$$y(n) - 0.5y(n-1) + 0.06y(n-2) = (0.4)^{n-1}u(n-1).$$

Determine the solution when the initial conditions are y(- 1) = 1 and y(- 2) = 2.

5.7. A system is described by the difference equation

$$y(n) - 0.5y(n-1) + 0.06y(n-2) = (0.4)^{n-1}u(n-1).$$

Determine the solution when the initial conditions are y(-1) = 1 and y(- 2) = 2.

5.8. Given the following difference equation with the input-output relationship of a certain initially relaxed system (all initial conditions are zero),

$$y(n) - 0.7y(n-1) + 0.1y(n-2) = x(n) + x(n-1),$$

   a. find the impulse response sequence y(n) due to the impulse sequence d(n);
   b. find the output response of the system when the unit step function u(n) is applied.

5.9. Given the following difference equation with the input-output relationship of a certain initially relaxed DSP system (all initial conditions are zero),

$$y(n) - 0.4y(n-1) + 0.29y(n-2) = x(n) + 0.5x(n-1),$$

   a. find the impulse response sequence $y(n)$ due to an impulse sequence $h(n)$;
   b. find the output response of the system when a unit step function $u(n)$ is applied.

Chapter Six

Frequency Analysis of Discrete Signals and Systems

Learning Outcomes of this Chapter

After successful completion of this chapter students will be able to:

1. learn techniques for representing discrete-time periodic signals using orthogonal sets of periodic basis functions.
2. study properties of exponential, trigonometric and compact Fourier series, and conditions for their existence.
3. learn the Fourier transform for non-periodic signal as an extension of Fourier series for periodic signals.
4. study the properties of the Fourier transform. Understand the concepts of energy and power spectral density.

6.1 Introduction

Although the time domain is the most natural, since everything (including our own lives) evolves in time, it is not the only possible representation. In most cases we want to know the frequency content of our signal. Most popular analysis in frequency domain is based on work of Joseph Fourier. The Fourier transform is one of several mathematical tools that is useful in the analysis and design of LTI systems. Another is the

Fourier series. These signal representations basically involve the decomposition of the signals in terms of sinusoidal (or complex exponential) components. With such a decomposition, a signal is said to be represented in the *frequency domain*.

Will begin with the frequency analysis of signals with the representation of continuous-time periodic and aperiodic signals by means of the Fourier series and the Fourier transform, respectively. This is followed by a parallel treatment of discrete time periodic and aperiodic signals. The properties of the Fourier transform are described in detail and a number of time-frequency dualities are presented.

6.2 Frequency analysis of a Continuous Time signal

Frequency analysis of a signal involves the resolution of the signal into its frequency (sinusoidal) components. Where these signal waveforms are basically functions of time. The role of the prism is played by the Fourier analysis tools that we will develop: the Fourier series and the Fourier transform. The recombination of the sinusoidal components to reconstruct the original signal is basically a Fourier synthesis problem. The spectrum provides an "identity" or a signature for the signal in the sense that no other signal has the same spectrum. As we will see, that attribute is related to the mathematical treatment of frequency-domain technique: If we decompose a waveform into sinusoidal components, in much the same Way that a prism separates white light into different colors, the sum of these sinusoidal components results in the original waveform. On the other hand, if any of these components is missing, the result is a different signal.

The basic motivation for developing the frequency analysis tools is to provide a mathematical and pictorial representation for the frequency components that are contained in any given signal. As in physics, the term *spectrum* is used when referring to the frequency content of a signal. The process of obtaining the spectrum of a given signal using the basic mathematical tools described in this chapter is known as *frequency* or *spectral analysis*. In contrast, the process of determining the spectrum of a signal in practice, based on actual measurements of the signal, is called *spectrum estimation*. This distinction is very important. In a practical problem the signal to be analyzed does not lend itself to an exact mathematical description. The signal is usually some information-bearing signal from which we are attempting to extract the relevant information. If the information that we wish to extract can be obtained either directly or indirectly from the spectral content of the signal, we can perform *spectrum estimation* the information-bearing signal, and thus obtain an estimate of the signal spectrum. In fact, we can view spectral estimation as a type of spectral analysis performed on signals obtained from physical sources (e.g., speech, EEG, ECG, etc.).

6.2.1 Fourier Series for Continuous-Time Periodic Signals

In this section we present the frequency analysis tools for continuous-time periodic signals. Examples of periodic signals encountered in practice are square waves, rectangular waves, triangular waves, and of course, sinusoids and complex exponentials. The basic mathematical representation of periodic signals is the Fourier series, which is a linear weighted sum of harmonically related sinusoids or complex exponentials. Jean Baptiste Joseph Fourier (1768-1830), a French mathematician, used such trigonometric series expansions in describing the phenomenon of heat conduction and temperature distribution through bodies. Although his work was motivated by the problem of heat conduction, the mathematical techniques that he developed during the early part of the nineteenth century now find application in a variety of problems encompassing many different fields, including optics, vibrations in mechanical systems, system theory, and electromagnetics.

We know that a linear combination of harmonically related complex exponentials of the form

$$x(t) = \sum_{k=-\infty}^{\infty} c_k e^{j2\pi k F_0 t}$$

(6.1)

is a periodic signal with fundamental period $T_p = 1/F_0$. Hence, we can think of the exponential signals

$$\{e^{j2\pi k F_0 t}, \quad k = 0, \pm 1, \pm 2, \ldots\}$$

Where F_0 determines the fundamental period of $x(t)$ and the coefficients specify the shape of the waveform.

Suppose that we are given a periodic signal $x(t)$ with period T_p. We can represent the periodic signal by the series (6.1), called a *Fourier series,* where the fundamental frequency F_0 is selected to be the reciprocal of the given period T_p. To determine the expression for the coefficients $\{C_k\}$, we first multiply both sides of (6.1) by the complex exponential.

$$e^{-j2\pi F_0 l t}$$

where l is an integer and then integrate both sides of the resulting equation over a single period, say from 0 to T_p, or more generally, from t_0 to $t_0 + T_p$, where t_0 is an arbitrary but mathematically convenient starting value. Thus we obtain

$$\int_{t_0}^{t_0+T_p} x(t)e^{-j2\pi l F_0 t}\, dt = \int_{t_0}^{t_0+T_p} e^{-j2\pi l F_0 t}\left(\sum_{k=-\infty}^{\infty} c_k e^{+j2\pi k F_0 t}\right) dt \tag{6.2}$$

To evaluate the integral on the right-hand side of (6.2), we interchange the order of the summation and integration and combine the two exponentials. Hence

$$\sum_{k=-\infty}^{\infty} c_k \int_{t_0}^{t_0+T_p} e^{j2\pi F_0(k-l)t}\, dt = \sum_{k=-\infty}^{\infty} c_k \left[\frac{e^{j2\pi F_0(k-l)t}}{j2\pi F_0(k-l)}\right]_{t_0}^{t_0+T_p} \tag{6.3}$$

For $k \neq l$, the right-hand side of (6.3) evaluated at the lower and upper limits, t_0 and $t_0 + T_p$ respectively, yields zero. On the other hand, if $k = l$, we have

$$\int_{t_0}^{t_0+T_p} dt = t\Big|_{t_0}^{t_0+T_p} = T_p$$

Consequently, (6.2) reduces to

$$\int_{t_0}^{t_0+T_p} x(t)e^{-j2\pi l F_0 t}\, dt = c_l T_p$$

and therefore the expression for the Fourier coefficients in terms of the given period' signal becomes

$$c_l = \frac{1}{T_p}\int_{t_0}^{t_0+T_p} x(t)e^{-j2\pi l F_0 t}\, dt$$

Since t_0 is arbitrary, this integral can be evaluated over any interval of length T_p, that is, over any interval equal to the period of the signal $x(t)$. Consequently, the integral for the Fourier series coefficients will be written as

$$c_l = \frac{1}{T_p} \int_{T_p} x(t) e^{-j2\pi l F_0 t} \, dt \qquad (6.4)$$

An important issue that arises in the representation of the periodic signal $x(t)$ by the Fourier series is whether or not the series converges to $x(t)$ for every value of t, that is, whether the signal $x(t)$ and its Fourier series representation

$$\sum_{k=-\infty}^{\infty} c_k e^{j2\pi k F_0 t} \qquad (6.5)$$

are equal at every value of t. The so-called *Dirichlet conditions* guarantee that the series (6.5) will be equal to $x(t)$, except at the values of t for which $x(t)$ is discontinuous. At these values of t, (6.5) converges to the midpoint (average value) of the discontinuity. The Dirichlet conditions are:

1. The signal $x(t)$ has a finite number of discontinuities in any period.
2. The signal $x(t)$ contains a finite number of maxima and minima during any period.
3. The signal $x(t)$ is absolutely integrable in any period, that is,

$$\int_{T_p} |x(t)| \, dt < \infty \qquad (6.6)$$

All periodic signals of practical interest satisfy these conditions.

The weaker condition, that the signal has finite energy in one period,

$$\int_{T_p} |x(t)|^2 \, dt < \infty \qquad (6.7)$$

guarantees that the energy in the difference signal

$$e(t) = x(t) - \sum_{k=-\infty}^{\infty} c_k e^{j2\pi k F_0 t}$$

is zero, although $x(t)$ and its Fourier series may not be equal for all values of t. Note that (6.6) implies (6.7), but not vice versa. Also, both (6.7) and the Dirichlet conditions are sufficient but not necessary conditions (i.e., there are signals that have a Fourier series representation but do not satisfy these conditions).

In summary, if $x(t)$ is periodic and satisfies the Dirichlet conditions, it can be represented in a Fourier series as in (6.1), where the coefficients are specified by (6.4).

Therefore, the Synthesis and Analysis equation of frequency Analysis of Continuous-Time Periodic Signal as follows:

$$x(t) = \sum_{k=-\infty}^{\infty} c_k e^{j2\pi k F_0 t} \qquad (6.8)$$

and the analysis

$$c_k = \frac{1}{T_p} \int_{T_p} x(t) e^{-j2\pi k F_0 t} \, dt \qquad (6.9)$$

In general, the Fourier coefficients C_k are complex valued. Moreover, it is easily shown that if the periodic signal is real, C_k and C_{-k} are complex conjugates. As a result, if

$$c_k = |c_k| e^{j\theta_k}$$

Then

$$c_{-k} = |c_k|^{-j\theta_k}$$

Consequently, the Fourier series may also be represented in the form

$$x(t) = c_0 + 2 \sum_{k=1}^{\infty} |c_k| \cos(2\pi k F_0 t + \theta_k)$$
(6.10)

where C_0 is real valued when $x(t)$ is real.

Finally, we should indicate that yet another form for the Fourier series can be obtained by expanding the cosine function in (6.10) as

$$\cos(2\pi k F_0 t + \theta_k) = \cos 2\pi k F_0 t \cos \theta_k - \sin 2\pi k F_0 t \sin \theta_k$$

Consequently, we can rewrite (6.10) in the form

$$x(t) = a_0 + \sum_{k=1}^{\infty} (a_k \cos 2\pi k F_0 t - b_k \sin 2\pi k F_0 t)$$
(6.11)

Where

$$a_0 = c_0$$
$$a_k = 2|c_k|\cos\theta_k$$
$$b_k = 2|c_k|\sin\theta_k$$

The expressions in (6.8), (6.10), and (6.11) constitute three equivalent forms for the Fourier series representation of a real periodic signal.

6.3 Frequency Analysis of Discrete-Time Signals

Discrete signals can be represented in the frequency domain by means of the Fourier Transform (FT), as seen for continuous signals. In the discrete case, the physical interpretation of the FT may be less evident, but nevertheless it is a very useful tool. In this section we repeat the development for the class of discrete-time signals. As mentioned in previous sections, the Fourier series representation of a continuous-time periodic signal can consist of an infinite number of frequency com ponents, where the frequency spacing between two successive harmonically related frequencies is $1/Tp$, and where Tp is the fundamental period. Since the frequency range for continuous-time signals extends from $-\infty$ to ∞, it is possible to have signals that contain an infinite number of frequency com ponents. In contrast, the frequency range for discrete-time signals is unique over the interval $(-\pi.\pi)$ or $(0, 2\pi)$. A discrete-time signal of fundamental period N can consist of frequency com ponents separated by $2n/N$ radians or $f = 1/N$ cycles. Consequently, the Fourier series representation of the discrete-tim e periodic signal will contain at m ost N frequency com ponents. This is the basic difference between the Fourier series representations for continuous-time and discrete-time periodic signals.

6.3.1 Fourier Series for Discrete-Time Periodic Signals

Suppose that we are given a periodic sequence $x(n)$ with period N. that is. $x(n) = x(n + N)$ for all N, The Fourier series representation for $x(n)$ consists of N harmonically related exponential functions

$$e^{j2\pi kn/N} \qquad k = 0, 1, \ldots, N-1$$

and is expressed as

$$x(n) = \sum_{k=0}^{N-1} c_k e^{j2\pi kn/N} \qquad (6.12)$$

where the $\{C_k\}$ are the coefficients in the series representation. To derive the expression for the Fourier coefficients, we use the following formula:

$$\sum_{n=0}^{N-1} e^{j2\pi kn/N} = \begin{cases} N, & k = 0, \pm N, \pm 2N, \ldots \\ 0, & \text{otherwise} \end{cases} \qquad (6.13)$$

Note the similarity of (6.13) with the continuous-time counterpart in (6.3). The proof of (6.13) follows immediately from the application of the geometric summation formula

$$\sum_{n=0}^{N-1} a^n = \begin{cases} N, & a = 1 \\ \dfrac{1-a^N}{1-a}, & a \neq 1 \end{cases} \qquad (6.14)$$

The expression for the Fourier coefficients C_k can be obtained by multiplying both sides of (6.12) by the exponential $e^{-j2\pi n/N}$ and summing the product from $n = 0$ to $n = N - 1$. Thus;

$$\sum_{n=0}^{N-1} x(n)e^{-j2\pi ln/N} = \sum_{n=0}^{N-1}\sum_{k=0}^{N-1} c_k e^{j2\pi(k-l)n/N} \qquad (6.15)$$

If we perform the summation over n first, in the right-hand side of (6.15), we obtain

$$\sum_{n=0}^{N-1} e^{j2\pi(k-l)n/N} = \begin{cases} N, & k-l = 0, \pm N, \pm 2N, \ldots \\ 0, & \text{otherwise} \end{cases} \qquad (6.16)$$

where we have made use of (6.13). Therefore, the right-hand side of (6.15) reduces to Nc_l and hence

$$c_l = \frac{1}{N}\sum_{n=0}^{N-1} x(n)e^{-j2\pi ln/N} \qquad l = 0, 1, \ldots, N-1 \qquad (6.17)$$

Thus, we have the desired expression for the Fourier coefficients in terms of the signal $x(n)$. The equations of analysis and synthesis of Frequency Analysis of Discrete-Time Periodic Signals as below:

Synthesis equation is:

$$x(n) = \sum_{k=0}^{N-1} c_k e^{j2\pi kn/N} \qquad (6.18)$$

Analysis equation is

$$c_k = \frac{1}{N}\sum_{n=0}^{N-1} x(n)e^{-j2\pi kn/N} \qquad (6.19)$$

The equation (6.18) is often called the *discrete-rime Fourier series* (DTFS). The Fourier coefficients $\{C_k\}$. $k = 0.1.....N-1$ provide the description of $x(n)$ in the frequency domain, in the sense that ck represents the amplitude and phase associated with the frequency component.

$$s_k(n) = e^{j2\pi kn/N} = e^{j\omega_k n}$$

where $\omega_k = 2\pi k / N$.

Note that the functions $s_k(n)$ are periodic with period N. Hence $s_t(n) = s_k(n+N)$. In view of this periodicity, it follows that the Fourier coefficients C_k when viewed beyond the range $k = 0,1....,N-1$, also satisfy a periodicity condition. Indeed, from (6.19), which holds for every value of k, we have

$$c_{k+N} = \frac{1}{N}\sum_{n=0}^{N-1} x(n)e^{-j2\pi(k+N)n/N} = \frac{1}{N}\sum_{n=0}^{N-1} x(n)e^{-j2\pi kn/N} = c_k \quad (6.20)$$

Therefore, the Fourier series coefficients $\{C_k\}$ form a periodic sequence when extended outside of the range $k = 0,1,....N-1$. Hence
$$C_k + N = C_k$$

that is, $\{C_k\}$ is a periodic sequence with fundamental period N. Thus the spectrum of a signal x(n), which is periodic with period N, is a periodic sequence with period N. Consequently, any N consecutive samples of the signal or its spectrum provide a complete description of the signal in the time or frequency domains.

Although the Fourier coefficients form a periodic sequence, we will focus our attention on the single period with range

$k = 0, 1, \ldots, N - 1$. This is convenient, since in the frequency domain, this amounts to covering the fundamental range $0 \leq \omega_k = \frac{2\pi k}{N} < 2\pi$, for $0 \leq k \leq N - 1$. In contrast, the frequency range $-\pi < \omega_k = \frac{2\pi k}{N} \leq \pi$, corresponds to $-N/2 < k \leq N/2$, which creates an inconvenience when N is odd. Clearly. if we use a sampling frequency F_s, the range $0 \leq k \leq N - 1$ corresponds to the frequency range $0 \leq F < F_s$,

Example 6.1
Determine the spectra of the signals

    a. $x(n) = \cos\sqrt{2}\,\pi n$
    b. $x(n) = \cos(\frac{\pi n}{3})$
    c. $x(n)$ is periodic with period $N = 4$ and
$$x(n) = \{\underset{\uparrow}{1}, 1, 0, 0\}$$

Solution
For $\omega_0 = \sqrt{2}\,\pi$, we have $f_0 = 1/\sqrt{2}$. Since f_0 is not a rational number, the signal is not periodic. Consequently. this signal cannot be expanded in a Fourier series. Nevertheless, the signal does possess a spectrum. Its spectral content consists of the single frequency component at $\omega = \omega_0 = \sqrt{2}\,\pi$.

In this case $f_0 = \frac{1}{6}$ and hence $x(n)$ is periodic with fundamental period $N = 6$.

From (6.19) we have

$$c_k = \frac{1}{6}\sum_{n=0}^{5} x(n) e^{-j2\pi k n/6} \qquad k = 0, 1, \ldots, 5$$

However, $x(n)$ can be expressed as

$$x(n) = \cos\frac{2\pi n}{6} = \frac{1}{2}e^{j2\pi n/6} + \frac{1}{2}e^{-j2\pi n/6}$$

which is already in the form of the exponential Fourier series in (6.18). In comparing the two exponential terms in $x(n)$ with (6.18), it is apparent that $c_1 = \frac{1}{2}$. The second exponential in x(n) corresponds to the term $k = -1$ in (6.18) However, this term can also be written a

$$e^{-j2\pi n/6} = e^{j2\pi(5-6)n/6} = e^{j2\pi(5n)/6}$$

which means that $c_{-1} = c_5$. But this is consistent with (6.20), and our previous observation that the Fourier series coefficients form a periodic sequence of period N. Consequently, we conclude that:
$$c_1 = c_2 = c_3 = c_4 = 0$$
$$c_1 = \frac{1}{2}, \; c_5 = \frac{1}{2},$$

a. From (4.2.8), we have

$$c_k = \frac{1}{4}\sum_{n=0}^{3} x(n)e^{-j2\pi kn/4} \qquad k = 0, 1, 2, 3$$

Or

$$c_k = \frac{1}{4}(1 + e^{-j\pi k/2}) \qquad k = 0, 1, 2, 3$$

For $k = 0, 1, 2, 3$ we obtain

$$c_0 = \tfrac{1}{2} \qquad c_1 = \tfrac{1}{4}(1-j) \qquad c_2 = 0 \qquad c_3 = \tfrac{1}{4}(1+j)$$

The magnitude and phase spectra are

$$|c_0| = \frac{1}{2} \qquad |c_1| = \frac{\sqrt{2}}{4} \qquad |c_2| = 0 \qquad |c_3| = \frac{\sqrt{2}}{4}$$

$$\angle c_0 = 0 \qquad \angle c_1 = -\frac{\pi}{4} \qquad \angle c_2 = \text{undefined} \qquad \angle c_3 = \frac{\pi}{4}$$

Figure 4.10 illustrates the spectral content of the signals in (b) and (c).

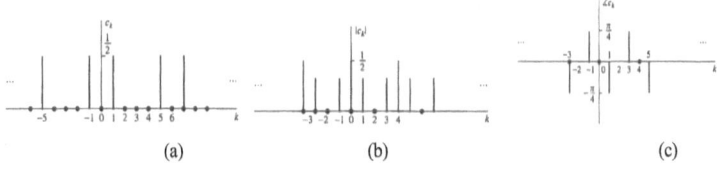

(a)   (b)   (c)

Figure 6.1 Spectra of the periodic signals discussed in Example 6.1 (b) and (c).

6.3.2 Fourier Transform of Discrete-Time Aperiodic Signals

Just as in the case of continuous-time aperiodic energy signals, the frequency analysis of discrete-time aperiodic finite-energy signals involves a Fourier transform of the time-domain signal.

The Fourier transform of a finite-energy discrete-time signal *x(n)* is defined as

$$X(\omega) = \sum_{n=-\infty}^{\infty} x(n) e^{-j\omega n} \tag{6.21}$$

Physically, $X(\omega)$ represents the frequency content of the signal $x(n)$. In other words, $X(\omega)$ is a decomposition of $x(n)$ into its frequency components.

We observe two basic differences between the Fourier transform of a discrete time finite-energy signal and the Fourier transform of a finite-energy analog signal. First, for continuous-time signals, the Fourier transform, and hence the spectrum of the signal, have a frequency range of $(-\infty, \infty)$. In contrast, the frequency range for a discrete-time signal is unique over the frequency interval of $(-\pi, \pi)$ or equivalently, $(0, 2\pi)$.

This property is reflected in the Fourier transform of the signal. Indeed. $X(\omega)$ is periodic with period 2π that is.

$$\begin{aligned} X(\omega + 2\pi k) &= \sum_{n=-\infty}^{\infty} x(n) e^{-j(\omega + 2\pi k)n} \\ &= \sum_{n=-\infty}^{\infty} x(n) e^{-j\omega n} e^{-j2\pi kn} \\ &= \sum_{n=-\infty}^{\infty} x(n) e^{-j\omega n} = X(\omega) \end{aligned} \qquad (6.22)$$

Hence $X(\omega)$ is periodic with period 2π. But this property is just a consequence of the fact that the frequency range for any discrete-time signal is limited to $(-\pi, \pi)$ or $(0, 2\pi)$, and any frequency outside this interval is equivalent to a frequency within the interval.

The second basic difference is also a consequence of the discrete-time nature of the signal. Since the signal is discrete in time. the Fourier transform of the signal involves a summation of terms instead of an integral, as in the case of continuous-time signals.

Since $X(\omega)$ is a periodic function of the frequency variable ω, it has a Fourier series expansion, provided that the conditions for the existence of the Fourier series. described previously, are satisfied. In fact, from the definition of the Fourier transform $X(\omega)$ of the sequence $x(n)$, given by (6.21), we observe that $X(\omega)$ has the form of a Fourier series. The Fourier coefficients in this series expansion are the values of the sequence $x(n)$.

To demonstrate this point. let us evaluate the sequence $x(n)$ from $X(\omega)$. First, we multiply both sides (6.21) by $e^{j\omega m}$ and integrate over the interval $(-\pi, \pi)$. Thus we have

$$\int_{-\pi}^{\pi} X(\omega)e^{j\omega m}d\omega = \int_{-\pi}^{\pi}\left[\sum_{n=-\infty}^{\infty} x(n)e^{-j\omega n}\right]e^{j\omega m}d\omega \tag{6.23}$$

The integral on the right-hand side of (6.23) can be evaluated if we can inter change the order of summation and integration. This interchange can be made if the series converges uniformly to $X(\omega)$ as $N - \infty$.

$$X_N(\omega) = \sum_{n=-N}^{N} x(n)e^{-j\omega n}$$

Uniform convergence means that, for every ω, $X_N(\omega) \to X(\omega)$, as $N - \infty$.. The convergence of the Fourier transform is discussed in more detail in the following section. For the moment, let us assume that the series converges uniformly, so that we can interchange the order of summation and integration in (6.23). Then

$$\int_{-\pi}^{\pi} e^{j\omega(m-n)} d\omega = \begin{cases} 2\pi, & m = n \\ 0, & m \neq n \end{cases}$$

Consequently,

$$\sum_{n=-\infty}^{\infty} x(n) \int_{-\pi}^{\pi} e^{j\omega(m-n)} d\omega = \begin{cases} 2\pi x(m), & m = n \\ 0, & m \neq n \end{cases} \quad (6.24)$$

By combining (6.23) and (6.24), we obtain the desired result that

$$x(n) = \frac{1}{2\pi} \int_{-\infty}^{\infty} X(\omega) e^{j\omega n} d\omega \quad (6.25)$$

If we compare the integral in (6.25) with (6.9), we note that this is just the expression for the Fourier series coefficient for a function that is periodic with period 2rr. The only difference between (6.9) and (6.25) is the sign on the exponent in the integrand. which is a consequence of our definition of the Fourier transform as given by (6.21). Therefore. the Fourier transform of the sequence $x(n)$, defined by (6.21), has the form of a Fourier series expansion.

6.4 Frequency Domain Representation of Discrete-time LTI Systems

It is customary to describe an LTI discrete-time system in terms of a linear difference equation with constant coefficients. To describe such a system in the frequency domain, we must determine its transfer function in the frequency domain. Consider an LTI discrete-time system described by the following difference equation, where the coefficients are constants:

$$y[n] = \sum_{j=0}^{q} a_j x[n-j] - \sum_{k=1}^{p} b_k y[n-k], p > q \quad (6.26)$$

By applying the DTFT on both sides of (6.26) and making use of the time shifting property of the DTFT, we get

$$Y(e^{j\Omega}) = X(e^{j\Omega}) \sum_{m=0}^{q} a_m e^{-jm\Omega} - Y(e^{j\Omega}) \sum_{n=1}^{p} b_n e^{-jn\Omega} \quad (6.27)$$

We can then express the transfer function of the LTI discrete-time system in the frequency domain as

$$H(e^{j\Omega}) = \frac{Y(e^{j\Omega})}{X(e^{j\Omega})} = \frac{a_0 + a_1 e^{-j\Omega} + a_2 e^{-j2\Omega} + \cdots + a_q e^{-jq\Omega}}{1 + b_1 e^{-j\Omega} + b_2 e^{-j2\Omega} + \cdots + b_p e^{-jp\Omega}} \quad (6.28)$$

As expected, the transfer function of an LTI discrete-time system in the frequency domain is a rational function in the variable $e^{j\Omega}$. Recall that the system described by (6.26) in the discrete-time domain or equivalently by (6.28) in the frequency domain is a recursive system. Hence its transfer function is a rational polynomial in the variable $e^{j\Omega}$. On the other hand, a non-recursive system will have a transfer function where the denominator is identically equal to 1.

Example 6.2 The impulse response of an LTI discrete-time system is given by $h[n] = 0.5^n u[n]$. If the input to this system is $x[n] = 0.75^n u[n]$, find the response of the system using the DTFT.

Solution: Since the given system is LTI, its response is the convolution of its impulse response sequence and the input sequence. In the frequency domain, the DTFT of the system response is the product of the DTFT of its impulse response and the input. First, we find the DTFT of the impulse response and the input from our previous discussion. Therefore,

$$H(e^{j\Omega}) = DTFT\{h[n]\} = \frac{1}{1-0.5e^{-j\Omega}} \tag{6.29a}$$

$$X(e^{j\Omega}) = DTFT\{x[n]\} = \frac{1}{1-0.75e^{-j\Omega}} \tag{6.29b}$$

The DTFT of the system output is

$$Y(e^{j\Omega}) = DTFT\{y[n]\} = DTFT\{h[n] \otimes x[n]\} = H(e^{j\Omega})X(e^{j\Omega}) = \frac{1}{(1-0.5e^{-j\Omega})(1-0.75e^{-j\Omega})} \tag{6.30}$$

The time domain response is the IDTFT of (6.30). One way to find the IDTFT is to express (6.30) in partial fractions and then identify each fraction with a real exponential sequence. So,

$$Y(e^{j\Omega}) = \frac{A}{1-0.5e^{-j\Omega}} + \frac{B}{1-0.75e^{-j\Omega}} \tag{6.31}$$

The residues are given by

$$A = (1-0.5e^{-j\Omega})Y(e^{j\Omega})\big|_{e^{j\Omega}=0.5} = -2 \tag{6.32a}$$

$$B = (1-0.75e^{-j\Omega})Y(e^{j\Omega})\big|_{e^{j\Omega}=0.75} = 3 \tag{6.32b}$$

Note that $IDTFT\left\{\frac{1}{1-0.5e^{-j\Omega}}\right\} = 0.5^n u[n]$ and $IDTFT\left\{\frac{1}{1-0.75e^{-j\Omega}}\right\} = 0.75^n u[n]$. Therefore, the response of the given LTI discrete-time system is

$$y[n] = -2(0.5)^n u[n] + 3(0.75)^n u[n] \tag{6.33}$$

Example 6.3 An LTI discrete-time system consists of a cascade of two systems $h_1[n]$ and $h_2[n]$, where,

$$h_1[n] = \delta[n] + \delta[n-1] \text{ and} \tag{6.34a}$$

$$h_2[n] = \beta^n u[n], |\beta| < 1 \tag{6.34b}$$

Determine the value of β such that the magnitude of the overall frequency response of the system is unity.

Solution: In a cascade connection, the output of the first system is the input to the second system and so on. This implies that the overall impulse response is the convolution of the impulse response of the individual sections. Therefore, the DTFT of the overall system is the product of the individual DTFTs. Thus,

$$H(e^{j\Omega}) = H_1(e^{j\Omega})H_2(e^{j\Omega}) = \frac{(1+e^{-j\Omega})}{1-\beta e^{-j\Omega}} \tag{6.35}$$

The magnitude of the overall frequency response is given by

$$|H(e^{j\Omega})| = \left|\frac{(1+\cos\Omega)-j\sin\Omega}{(1-\beta\cos\Omega)+j\beta\sin\Omega}\right| = \sqrt{\frac{2+2\cos\Omega}{1+\beta^2-2\beta\cos\Omega}} \Rightarrow \beta = -1 \tag{6.36}$$

A discrete-time LTI system whose magnitude of the frequency response is a constant is known as an *allpass* system.

6.4.1 Steady State Response of LTI Discrete-time Systems

We have seen earlier that the particular solution of a linear difference equation with constant coefficients is proportional to the input sequence. The complementary solution is in general, a decaying function. So, when the complementary solution or the transient response disappears, only the particular solution remains. This is called the *steady state* response of the system. In particular, when the input to an LTI discrete-time system is a sinusoidal sequence of a specified frequency, its response in the steady state is the same input sinusoid except that its amplitude and phase are modified by the value of the transfer function at that frequency. We can, therefore, express the steady state response of an LTI discrete-time system to an input sinusoid $e^{jn\Omega_0}$ as

$$y_{ss}[n] = |H(e^{j\Omega_0})| e^{j\Omega_0(n-\tau)} \qquad (6.37)$$

From equation (4.37) we notice that the amplitude of the output sinusoid is the magnitude of the transfer function at the input frequency and the lagging phase angle equals τ times the input frequency. If the phase response is linear, then the phase delay equals the negative of the phase angle divided by the input frequency. If the phase angle represented by $\theta(\Omega)$ is linear, then the phase delay or simply the delay is expressed by

$$\tau = -\frac{\theta(\Omega)}{\Omega_0} \text{ samples} \qquad (6.38)$$

Recall that the transfer function $H(e^{j\Omega})$ is the DTFT of the impulse response $h[n]$ of an LTI discrete-time system. Since the impulse response is unique to a given system, the corresponding frequency response is also unique to the system.

Example 6.4 The impulse response of a discrete-time LTI system is given by

$$h[n] = 0.5^n u[n] \qquad (6.39)$$

Find the steady state response of the system if the input is $x[n] = \cos\left(\frac{n\pi}{6}\right) u[n]$.

Solution: The frequency response or the transfer function of the given system is the DTFT of its impulse response:

$$H(e^{j\Omega}) = DTFT\{h[n]\} = DTFT\{0.5^n u[n]\} = \frac{1}{1 - 0.5e^{-j\Omega}} \qquad (6.40)$$

The transfer function in magnitude-phase form of the given system is found to be

$$|H(e^{j\Omega})| = \frac{1}{\sqrt{1.25 - \cos\Omega}}, \qquad (6.41a)$$

$$\theta(\Omega) = -\tan^{-1}\left(\frac{0.5\sin\Omega}{1 - 0.5\cos\Omega}\right), rad \qquad (6.41b)$$

Since the input is a unit amplitude sinusoid at a frequency $\frac{\pi}{6}$ rad, the steady state response of the given system is given by

$$y_{ss}[n] = \left|H\left(e^{j\frac{\pi}{6}}\right)\right| \cos\left(\frac{n\pi}{6} - \theta\left(\frac{\pi}{6}\right)\right) u[n] \qquad (6.42)$$

From equations (4.41a) and (4.41b), we find that $\left|H\left(e^{j\frac{\pi}{6}}\right)\right| \approx 1.6138$ and the phase angle is $\theta\left(\frac{\pi}{6}\right) \approx -0.415283^r$ or $-23.79°$. Therefore, the steady state response of the given system to the input sinusoid is

$$y_{ss}[n] = 1.6138 \cos\left(\frac{n\pi}{6} - 23.79°\right) u[n] \qquad (6.43)$$

The input and the system response are plotted as a function of the sample index and shown in Figure 6.2 as top and bottom plots, respectively. The system response is obtained by calling the function *conv*. It accepts two sequences as vectors and returns a sequence that is of length equal to the sum of the lengths of the impulse response and input minus one. However, we plot the response in length equal to the input sequence. As can be seen from the figure, the system response is the same sinusoid as the input with a change in its amplitude. We also notice a delay of 1 sample in the output sequence due to the phase shift in the transfer function, which agrees with the analytical result.

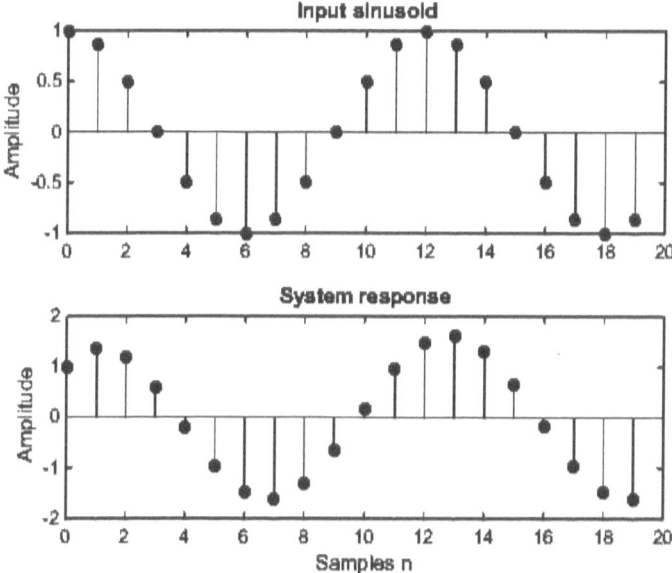

Figure 6.2 Steady state response of the system in Example 6.9 due to a sinusoidal input: Top plot: input sequence, Bottom plot: output sequence

Group delay: When the phase response of an LTI discrete-time system is not linear, then its phase delay is not constant but is a function of the input frequency. When a group of sinusoidal frequencies is present in the input, it is customary to find the phase delay over this group of frequencies. It is called the *group delay* and is defined as

$$\tau_{gd} = -\frac{d\theta(\Omega)}{d\Omega} \tag{6.60}$$

Let us illustrate the calculation of group delay by way of an example.

Example 6.5 Calculate the group delay of the system in Example 6.9 and plot the result.

Solution: The phase response of the given system is shown in equation (6.57b). Then, the group delay is given by

$$\tau_{gd} = -\frac{d}{d\Omega}\left\{-\tan^{-1}\left(\frac{0.5\sin\Omega}{1-0.5\cos\Omega}\right)\right\} \qquad (6.61)$$

To obtain the derivative in equation (6.61), first let us rewrite (6.57b) as

$$\tan\theta(\Omega) = \frac{-0.5\sin\Omega}{1-0.5\cos\Omega} \qquad (6.62)$$

Now differentiate (6.62) with respect to Ω. Therefore,

$$\frac{1}{\cos^2\theta}\frac{d\theta}{d\Omega} = \frac{-(1-0.5\cos\Omega)(0.5\cos\Omega)+(0.5\sin\Omega)(0.5\sin\Omega)}{(1-0.5\cos\Omega)^2} \qquad (6.63)$$

From (6.62), we get

$$\cos\theta(\Omega) = \frac{1-0.5\cos\Omega}{\sqrt{1.25-\cos\Omega}}. \qquad (6.64)$$

Using (6.64) in (6.63) and after algebraic manipulation, we obtain

$$\frac{d\theta(\Omega)}{d\Omega} = \frac{0.25-0.5\cos\Omega}{1.25-\cos\Omega} \qquad (6.65)$$

Using equation (6.65) in (6.61), we finally obtain the expression for the group delay as

$$\tau_{gd} = \frac{-(0.25-0.5\cos\Omega)}{1.25-\cos\Omega} \qquad (6.66)$$

The group delay in (6.66) is shown in Figure 6.3 top plot as a function Ω in the interval $0 \leq \Omega \leq \pi$. It accepts the coefficients of the numerator and denominator polynomials of the transfer function, both in ascending powers of $e^{-j\Omega}$. The group delay is shown in the bottom plot in Figure 6.3.

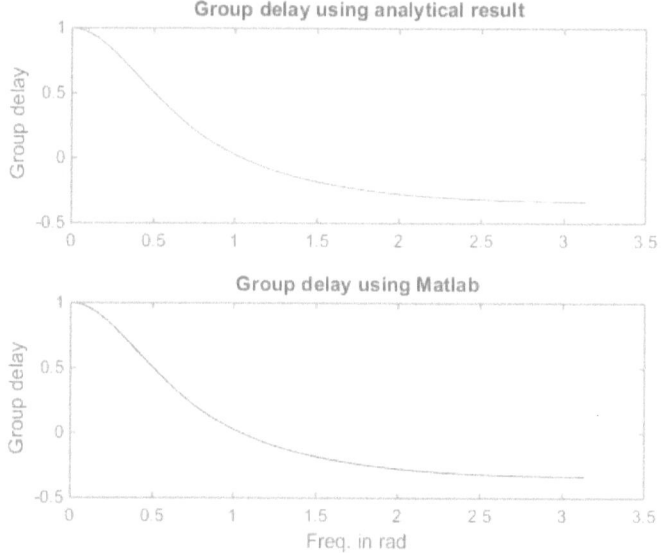

Figure 6.3 Group delay of the system in Example 6.9: Top plot: group delay obtained from equation (6.66),

6.5 Frequency Response of Systems

Systems are analyzed in the *time domain* by using convolution. A similar analysis can be done in the *frequency domain*. Using the Fourier transform, every input signal can be represented as a group of cosine waves, each with a specified amplitude and phase shift. Likewise, the DFT can be used to represent every output signal in a similar form. This means that any linear system can be *completely* described by how it changes the amplitude and phase of cosine waves passing through it. This information is called the system's **frequency response.** Since both the impulse response and the frequency response contain complete information about the system, there must be a one- to-one correspondence between the two. Given one, you can calculate the other. The relationship between the impulse response and the frequency response is one of the foundations of signal processing: *A system's frequency response is the Fourier Transform of its impulse response.* Figure 6-4 illustrates these relationships.

Keeping with standard DSP notation, impulse responses use lower case variables, while the corresponding frequency responses are upper case. Since $h[.]$ is the common symbol for the impulse response, $H[.]$ is used for the frequency response. Systems are described in the time domain by convolution, that is:

$x[n]*h[n] = y[n]$. In the frequency domain, the input spectrum is *multiplied* by the frequency response, resulting in the output spectrum. As an equation: $X[f] \times H[f] = Y[f]$. That is, *convolution* in the time domain corresponds to *multiplication* in the frequency domain.

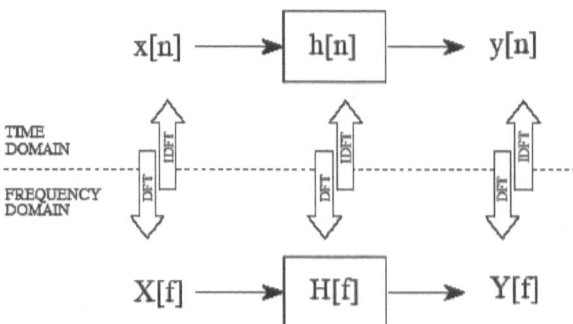

Figure 6.4. Comparing system operation in the time and frequency domains.

Figure 6.5 shows an example of using the DFT to convert a system's impulse response into its frequency response. Figure (a) is the impulse response of the system. Looking at this curve isn't going to give you the slightest idea what the system does. Taking a 64 point DFT of this impulse response produces the frequency response of the system, shown in (b). Now the function of this system becomes obvious, it passes frequencies between 0.2 and 0.3, and rejects all others. It is a band-pass filter. The *phase* of the frequency response could also be examined; however, it is *more* difficult to interpret and *less* interesting. Figure 6.5 (b) is very jagged due to the low number of samples defining the curve. This situation can be improved by **padding** the impulse response with zeros before taking the DFT. For example, adding zeros to make the impulse response 512 samples long, as shown in (c), results in the higher resolution frequency response shown in (d).

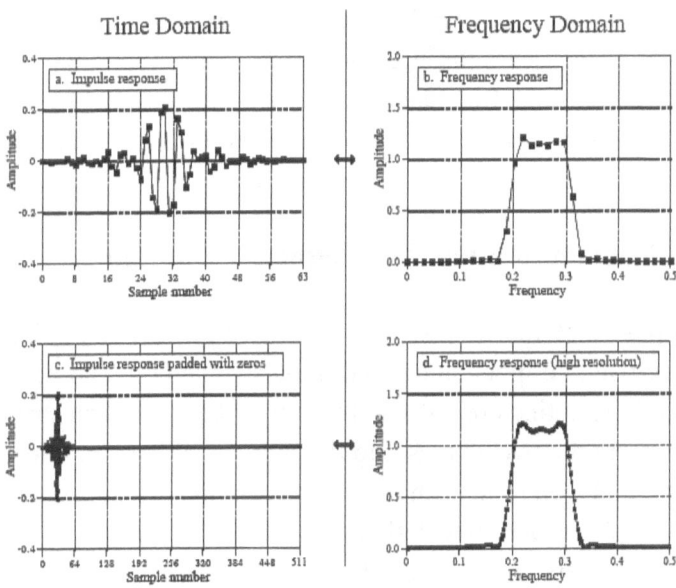

Figure 6.5. Finding the frequency response from the impulse response. By using the DFT, a system's impulse response, (a), can be transformed into the system's frequency response, (b). By adding the impulse response with zeros (c) higher resolution cab be obtained in the frequency response, (d). Only the magnitude of the frequency response is

How much resolution can you obtain in the frequency response? The answer is: *infinitely* high, if you are willing to pad the impulse response with an *infinite* number of zeros. In other words, there is nothing limiting the frequency resolution except the length of the DFT. This leads to a very important concept. Even though the impulse response is a *discrete* signal, the corresponding frequency response is *continuous*. An N point DFT of the impulse response provides $N/2 + 1$ *samples* of this continuous curve. If you make the DFT longer, the resolution improves, and you obtain a better idea of what the continuous curve looks like. Remember what the frequency response represents: amplitude and phase changes experienced by cosine waves as they pass through the system. Since the input signal can contain any frequency between 0 and 0.5, the system's frequency response must be a continuous curve over this range.

6.6 Convolution via the Frequency Domain

Suppose that you despise convolution. What are you going to do if given an input signal and impulse response, and need to find the resulting output signal? Figure 9-8 provides an answer: transform the two signals into the frequency domain, multiply them, and then transform the result back into the time domain. This replaces one convolution with two DFTs, a multiplication, and an Inverse DFT. Even though the intermediate steps are very different, the output is *identical* to the standard convolution algorithm.

Does anyone hate convolution enough to go to this trouble? The answer is yes. Convolution is avoided for two reasons. First, convolution is *mathematically* difficult to deal with. For instance, suppose you are given a system's impulse response, and its output signal. How do you calculate what the input signal is? This is called deconvolution and is virtually impossible to understand in the time domain. However, deconvolution can be carried out in the frequency domain as a simple *division*, the inverse operation of multiplication. The frequency domain becomes attractive whenever the complexity of the Fourier Transform is less than the complexity of the convolution. This isn't a matter of which you like better; it is a matter of which you hate less. The second reason for avoiding convolution is *computation speed*. For example, suppose you design a digital filter with a kernel (impulse response) containing 512 samples. Using a 200 MHz personal computer with floating point numbers, each sample in the output signal requires about one millisecond to calculate, using the standard convolution algorithm. In other words, the throughput of the system is only about 1,000 samples per second. This is 40 times too slow for high-fidelity audio, and 10,000 times too slow for television quality video!

The standard convolution algorithm is slow because of the large number of multiplications and additions that must be calculated. Unfortunately, simply bringing the problem into the frequency domain via the DFT doesn't help at all. Just as many calculations are required to calculate the DFTs, as are required to directly calculate the convolution. A breakthrough was made in the problem in the early 1960s when the *Fast Fourier Transform* (FFT) was developed.

Problem

6.1. Compute the complex-form Fourier series coefficients and sketch the magnitude and phase spectra for

(a) the signal $x(t)$ that has fundamental period $T_0 = 1$, with $x(t) = e^{-t}, 0 \leq t \leq 1$.

(b) the signal
$$x(t) = \sum_{k=-\infty}^{\infty} (-1)^k \delta(t - 2k)$$

(c) the signal $x(t)$ shown below

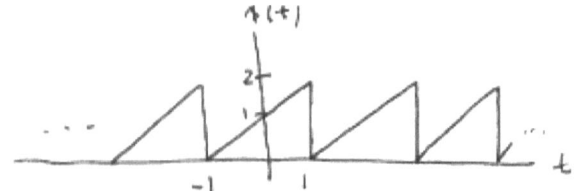

6.2 Suppose $x(t)$ is periodic with fundamental period T_0 and complex-form Fourier series coefficients X_k. Show that

(a) if $x(t)$ is odd, $x(t) = -x(-t)$, then $X_k = -X_{-k}$ for all k.

(b) if $x(t)$ is "half-wave odd," $x(t) = -x(t + T_0/2)$, then $X_k = 0$ for every even integer k.

(c) if $x(t)$ is even, $x(t) = x(-t)$, then $X_k = X_{-k}$ for all k.

6.3. Given the LTI system with unit-impulse response $h(t) = e^{-4|t|}$, compute the Fourier series representation for the response $y(t)$ of the system to the input signal

(a) $x(t) = \sum_{n=-\infty}^{\infty} \delta(t - n)$

(b) $x(t) = \sum_{n=-\infty}^{\infty} (-1)^n \delta(t - n)$

6.4. A continuous-time periodic signal $x(t)$ has Fourier series coefficients

$$X_k = \begin{cases} (6/jk)e^{jk\pi/4}, & k = \pm 1, \pm 3 \\ 0, & else \end{cases}$$

Compute and sketch the magnitude and phase spectra of the signal

6.5. Compute the discrete-time Fourier series coefficients for the signals below and sketch the magnitude and phase spectra.

(a) $x[n] = 1 + \cos(\pi n/3)$
(b) $[n] = \sum_{k=-\infty}^{\infty} \delta(n - 4k - 1)$

6.6. For the sets of DTFS coefficients given below, determine the corresponding real, periodic signal $x[n]$.

(a) $X_k = \begin{cases} \frac{1}{2}, & k \text{ even,} \\ -\frac{1}{2}, & k \text{ odd} \end{cases} \quad \omega_0 = \pi$

(b) $X_k = \{\frac{1}{2}, \quad \text{for all } k, \quad \omega_0 = \pi$
(c) $X_0 = -1, \ X_1 = 0, \ X_2 = 1, \ X_3 = -2, \ X_4 = 1, \ X_5 = 0, \ X_{k+6} = X_k, \ \omega_0 = \pi/3$

6.7. Suppose $x[n]$ is periodic with *even* fundamental period N_0 and DTFS coefficients X_k. If $x[n]$ also satisfies $x[n] = -x[n + N_0/2]$, for all n, show that $X_k = 0$ if k is even.

6.8. If $x[n]$ has fundamental period N_0, an even integer, and discrete-time Fourier series coefficients X_k, what are the Fourier series coefficients for:

(a) $\hat{x}[n] = x[n + \frac{N_0}{2}]$

(b) $\hat{x}[n] = (-1)^n x[n]$ (Assume that $\widehat{N_0} = N_0$ and give an example to show why this assumption is needed.)

6.9. For the LTI systems specified below, sketch the magnitude of the frequency response function and determine if the system is a low-pass or high-pass filter.

(a) $h[n] = \frac{1}{2}\delta[n] - \frac{1}{2}\delta[n-1]$
(b) $h[n] = \delta[n] - (1/2)^n u[n]$
(c) $h[n] = (1/2)^n u[n]$

Chapter Seven

Discrete Fourier Transform

Learning Outcomes of this Chapter

After successful completion of this chapter students will be able to:

1. understanding the relationships between the transform, discrete-time Fourier transform (DTFT), discrete Fourier series (DFS), discrete Fourier transform (DFT) and fast Fourier transform (FFT).
2. understand the characteristics and properties of DFS and DFT.
3. perform discrete-time signal conversion between the time and frequency domains using DFS and DFT and their inverse transforms.

7.1 Introduction

In time domain, representation of digital signals describes the signal amplitude versus the sampling time instant or the sample number. However, in some applications, signal frequency content is very useful otherwise than as digital signal samples. The representation of the digital signal in terms of its frequency component in a frequency domain, that is, the signal spectrum, needs to be developed. Therefore, the discrete Fourier transform (DFT) is an equation for converting time domain data into frequency domain data. Discrete means that the signal is sampled in time rather than being continuous. Therefore, the DFT is an approximation for the continuous Fourier transform. Figure 7.1 illustrates the time domain representation of a 1,000-Hz sinusoid with 32 samples at a sampling rate of 8,000 Hz; the bottom plot shows the signal spectrum (frequency domain representation), where we can clearly observe that the amplitude peak is located at the frequency of 1,000 Hz in the calculated spectrum. Hence, the spectral plot better displays frequency information of a digital signal. The algorithm transforming the time domain signal samples to the frequency domain components is known as the discrete Fourier transform, or DFT. The DFT also establishes a relationship between the time domains representation

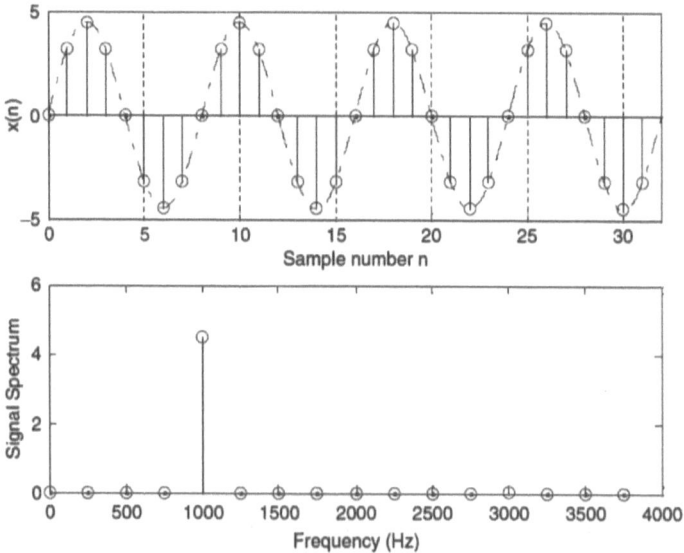

Figure 7.1 Example of the digital signal and its amplitude spectrum.

Discrete Fourier Transform (DFT) is the equivalent of the continuous Fourier Transform for signals known only at instants separated by sample times (i.e. a finite sequence of data). It is the family member used with digitized signals. it is It is a transformation that maps an N-point Discrete-time (DT) signal $x[n]$ into a function of the N complex discrete harmonics.

The general term: Fourier transform, can be broken into four categories, resulting from the four basic types of signals that can be encountered.

A signal can be either continuous or discrete, and it can be either periodic or aperiodic. Aperiodic-Continuous This includes, for example, decaying exponentials and the Gaussian curve. These signals extend to both positive and negative infinity without repeating in a periodic pattern. The Fourier Transform for this type of signal is simply called the Fourier Transform. Periodic-Continuous Here the examples include: sine waves, square waves, and any waveform that repeats itself in a regular pattern from negative to positive infinity. This version of the Fourier transform is called the Fourier Series. Aperiodic-Discrete These signals are only defined at discrete points between positive and negative infinity, and do not repeat themselves in a periodic fashion. This type of Fourier transform is called the Discrete Time Fourier Transform. Periodic-Discrete These are discrete signals that repeat themselves in a periodic fashion from negative to positive infinity. This class of Fourier Transform is sometimes called the Discrete Fourier Series, but is most often called the Discrete Fourier Transform.

Type of Fourier transform that can be used in DSP is the DFT. In other words, digital computers can only work with information that is discrete and finite in length. When you struggle with theoretical issues, grapple with homework problems, and ponder mathematical mysteries, you may find yourself using the first three members of the Fourier transform family. When you sit down to your computer, you will only use the DFT.

Let $f(t)$ be the continuous signal which is the source of the data. Let samples be denoted $f[0]$,
$$f[0], f[2], \ldots, f[k], \ldots, f[N-1]$$

The Fourier Transform of the original signal, $f(t)$, would be :
$$F(jw) = \int_{-\infty}^{\infty} f(t) e^{-jwt} dt \qquad (7.1)$$

We could regard each sample $f[k]$ as an impulse having area $f[k]$. Then, since the integrand exists only at the sample points:

$$F(jw) = \int_{0}^{(N-1)T} f(t) e^{-jwt} dt \qquad (7.2)$$
$$= f[0]e^{-j0} + f[1]e^{-jwT} + \ldots + f[k]e^{-jwkT} + \ldots + f[N-1]e^{-jw(N-1)T}$$

$$ei.e\ F(jw) = \sum_{k=0}^{N-1} f[k] e^{-jwkT}$$

We could in principle evaluate this for any #, but with only data points to start with, only final outputs will be significant. You may remember that the continuous Fourier transform could be evaluated over a finite interval (usually the fundamental period T_0) rather than from $-\infty$ to $+\infty$ if the waveform was *periodic*. Similarly, since there are only a finite number of input data points, the DFT treats the data as if it were periodic (i.e. to $f[N]$ to $f[2N-1]$ is the same as $f(0)$ to $f(N-1)$).

Hence the sequence shown below in Figure 7.2(a) is considers to be one period of the periodic sequence in figure 7.2 (b).

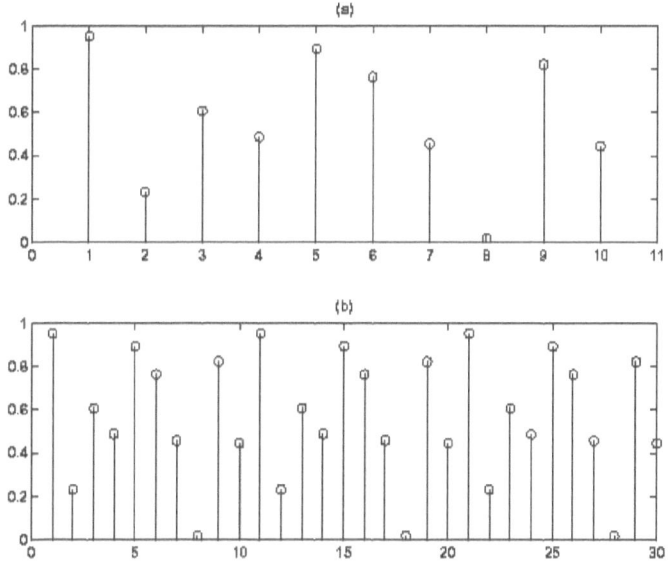

Figure 7.2 (a) Sequence of N=10 _ samples.
(b) implicit periodicity in DFT.

Therefore, the M-point DFT and inverse DFT for a time-limited sequence $x[k]$, which is non-zero within the limits $0 \le k \le (N-1)$, is given by

DFT synthesis equation

$$x[k] = \frac{1}{M}\sum_{r=0}^{M-1} X[r]e^{j\frac{2\pi rk}{M}} \quad \text{for } 0 \le k \le (M-1); \tag{7.3}$$

DFT analysis equation

$$X[r] = \sum_{k=0}^{M-1} x[k]e^{-j\frac{2\pi rk}{M}} \quad \text{for } 0 \le r \le (M-1). \tag{7.4}$$

By substituting the expression for $x[k]$ from the synthesis equation, Eq. (7.3), the analysis equation, Eq. (7.4), can be formally proved. In both equations, the length M of the DFT is typically set to be greater or equal to the length N of the aperiodic sequence $x[k]$. Unless otherwise stated, in case $M = N$ in the discussion that follows. Collectively, the DFT pair is denoted as

$$x[k] \xleftrightarrow{\text{DFT}} X[r]. \qquad (7.5)$$

Examples 7.1 and 7.2 illustrate the steps involved in calculating the DFTs of aperiodic sequences.

Example 7.1

Calculate the four-point DFT of the aperiodic sequence $x[k]$ of length $N = 4$, which is defined as follows:

$$x[k] = \begin{cases} 2 & k = 0 \\ 3 & k = 1 \\ -1 & k = 2 \\ 1 & k = 3. \end{cases}$$

Solution

The four-point DFT of $x[k]$ is given by

$$X[r] = \sum_{k=0}^{3} x[k] e^{-j(2\pi kr/4)}$$
$$= 2 + 3 \times e^{-j(2\pi r/4)} - 1 \times e^{-j(2\pi(2)r/4)} + 1 \times e^{-j(2\pi(3)r/4)},$$

for $0 \leq r \leq 3$. On substituting different values of r, we obtain

$r = 0$ $X[0] = 2 + 3 - 1 + 1 = 5$;

$r = 1$ $X[1] = 2 + 3 \times e^{-j(2\pi/4)} - 1 \times e^{-j(2\pi(2)/4)} + 1 \times e^{-j(2\pi(3)/4)}$

$= 2 + 3(-j) - 1(-1) + 1(j) = 3 - 2j$;

$r = 2$ $X[2] = 2 + 3 \times e^{-j(2\pi(2)/4)} - 1 \times e^{-j(2\pi(2)(2)/4)} + 1 \times e^{-j(2\pi(3)(2)/4)}$

$= 2 + 3(-1) - 1(1) + 1(-1) = -3$;

$r = 3$ $X[3] = 2 + 3 \times e^{-j(2\pi(3)/4)} - 1 \times e^{-j(2\pi(2)(3)/4)} + 1 \times e^{-j(2\pi(3)(3)/4)}$

$= 2 + 3(j) - 1(-1) + 1(-j) = 3 + j2$.

Example 7.2

Calculate the inverse DFT of
$$X[k] = [5, 3 - j2, -3, 3 + j2]$$

Solution

Using the inverse DFT of $X[r]$ is given by

$$x[k] = \frac{1}{4}\sum_{k=0}^{3} X[r]e^{j(2\pi kr/4)} = \frac{1}{4}\left[5 + (3 - j2) \times e^{j(2\pi k/4)} - 3 \times e^{j(2\pi(2)k/4)} + (3 + j2) \times e^{j(2\pi(3)k/4)}\right],$$

for $0 \leq k \leq 3$. On substituting different values of k, we obtain

$$x[0] = \frac{1}{4}[5 + (3 - j2) - 3 + (3 + j2)] = 2;$$

$$x[1] = \frac{1}{4}\left[5 + (3 - j2)e^{j(2\pi/4)} - 3e^{j(2\pi(2)/4)} + (3 + j2)e^{j(2\pi(3)/4)}\right]$$

$$= \frac{1}{4}[5 + (3 - j2)(j) - 3(-1) + (3 + j2)(-j)] = 3;$$

$$x[2] = \frac{1}{4}\left[5 + (3 - j2)e^{j(2\pi(2)/4)} - 3e^{j(2\pi(2)(2)/4)} + (3 + j2)e^{j(2\pi(3)(2)/4)}\right]$$

$$= \frac{1}{4}[5 + (3 - j2)(-1) - 3(1) + (3 + j2)(-1)] = -1;$$

$$x[3] = \frac{1}{4}\left[5 + (3 - j2)e^{j(2\pi(3)/4)} - 3e^{j(2\pi(2)(3)/4)} + (3 + j2)e^{j(2\pi(3)(3)/4)}\right]$$

$$= \frac{1}{4}[5 + (3 - j2)(-j) - 3(-1) + (3 + j2)(j)] = 1.$$

Examples 7.1 and 7.2 prove the following DFT pair:

$$x[k] = \begin{cases} 2 & k=0 \\ 3 & k=1 \\ -1 & k=2 \\ 1 & k=3 \end{cases} \xrightarrow{\text{DFT}} X[r] = \begin{cases} 5 & r=0 \\ 3-j2 & r=1 \\ -3 & r=2 \\ 3+j2 & r=3, \end{cases}$$

where both the DT sequence x[k] and its DFT $X[r]$ are aperiodic with length $N = 4$.

7.2 DFT as matrix multiplication

An alternative representation for computing the DFT is obtained by expanding Eq. (7.4) in terms of the time and frequency indices (k,r). For $N = M$, the resulting equations are expressed as follows:

$$\left.\begin{aligned}
X[0] &= x[0] + x[1] + x[2] + \cdots + x[N-1], \\
X[1] &= x[0] + x[1]e^{-j(2\pi/N)} + x[2]e^{-j(4\pi/N)} \\
&\quad + \cdots + x[N-1]e^{-j(2(N-1)\pi/N)}, \\
X[2] &= x[0] + x[1]e^{-j(4\pi/N)} + x[2]e^{-j(8\pi/N)} \\
&\quad + \cdots + x[N-1]e^{-j(4(N-1)\pi/N)}, \\
&\vdots \\
X[N-1] &= x[0] + x[1]e^{-j(2(N-1)\pi/N)} + x[2]e^{-j(4(N-1)\pi/N)} \\
&\quad + \cdots + x[N-1]e^{-j(2(N-1)(N-1)\pi/N)},
\end{aligned}\right\} \quad (7.6)$$

In the matrix-vector format they are given by

$$\begin{bmatrix} X[0] \\ X[1] \\ X[2] \\ \vdots \\ X[N-1] \end{bmatrix} = \begin{bmatrix} 1 & 1 & 1 & \cdots & 1 \\ 1 & e^{-j(2\pi/N)} & e^{-j(4\pi/N)} & \cdots & e^{-j(2(N-1)\pi/N)} \\ 1 & e^{-j(4\pi/N)} & e^{-j(8\pi/N)} & \cdots & e^{-j(4(N-1)\pi/N)} \\ \vdots & \vdots & \vdots & \ddots & \vdots \\ 1 & e^{-j(2(N-1)\pi/N)} & e^{-j(4(N-1)\pi/N)} & \cdots & e^{-j(2(N-1)(N-1)\pi/N)} \end{bmatrix} \begin{bmatrix} x[0] \\ x[1] \\ x[2] \\ \vdots \\ x[N-1] \end{bmatrix}.$$

$$\underbrace{}_{\text{DFT vector: } \tilde{X}} \quad \underbrace{}_{\text{DFT matrix: } F} \quad \underbrace{}_{\text{signal vector: } \tilde{x}}$$

(7.7)

Equation (7.7) shows that the DFT coefficients $X[r]$ can be computed by left- multiplying the DT sequence $x[k]$, arranged in a column vector $\tilde{x}$ in ascending order with respect to the time index k, by the DFT matrix F.

Similarly, the expression for the inverse DFT given in Eq. (7.3) can be expressed as follows:

$$\begin{bmatrix} x[0] \\ x[1] \\ x[2] \\ \vdots \\ x[N-1] \end{bmatrix} = \frac{1}{N} \begin{bmatrix} 1 & 1 & 1 & \cdots & 1 \\ 1 & e^{j(2\pi/N)} & e^{j(4\pi/N)} & \cdots & e^{j(2(N-1)\pi/N)} \\ 1 & e^{j(4\pi/N)} & e^{j(8\pi/N)} & \cdots & e^{j(4(N-1)\pi/N)} \\ \vdots & \vdots & \vdots & \ddots & \vdots \\ 1 & e^{j(2(N-1)\pi/N)} & e^{j(4(N-1)\pi/N)} & \cdots & e^{j(2(N-1)(N-1)\pi/N)} \end{bmatrix} \begin{bmatrix} X[0] \\ X[1] \\ X[2] \\ \vdots \\ X[N-1] \end{bmatrix},$$

$$\underbrace{}_{\text{signal vector: } x} \quad \underbrace{}_{\text{DFT matrix: } G=F^{-1}} \quad \underbrace{}_{\text{DFT vector: } x}$$

(7.8)

which implies that the DT sequence $x[k]$ can be obtained by left-multiplying the DFT coefficients $X[r]$, arranged in a column vector $\tilde{X}$ in ascending order with respect to the DFT coefficient index r, by the inverse DFT matrix G and then scaling the result by a factor $1/N$. It is straightforward to show that $G \times F = F \times G = N I_N$, where I_N is the identity matrix of order N. Example 7.3 repeats Example 7.1 using the matrix-vector representation for the DFT.

Example 7.3

Calculate the four-point DFT of the aperiodic signal $x[k]$ considered in Example 7.1.

Solution

Arranging the values of the DT sequence in the signal vector x, we obtain

$$x = [2 \quad 3 \quad -1 \quad 1]^T$$

where superscript T represents the transpose operation for a vector. Using Eq. (7.7), we obtain

$$\begin{bmatrix} X[0] \\ X[1] \\ X[2] \\ X[3] \end{bmatrix} = \underbrace{\begin{bmatrix} 1 & 1 & 1 & 1 \\ 1 & e^{-j(2\pi/N)} & e^{-j(4\pi/N)} & e^{-j(6\pi/N)} \\ 1 & e^{-j(4\pi/N)} & e^{-j(8\pi/N)} & e^{-j(12\pi/N)} \\ 1 & e^{-j(6\pi/N)} & e^{-j(12\pi/N)} & e^{-j(18\pi/N)} \end{bmatrix}}_{\text{DFT matrix: } F} \begin{bmatrix} x[0] \\ x[1] \\ x[2] \\ x[3] \end{bmatrix}$$

$$= \underbrace{\begin{bmatrix} 1 & 1 & 1 & 1 \\ 1 & e^{-j(2\pi/N)} & e^{-j(4\pi/N)} & e^{-j(6\pi/N)} \\ 1 & e^{-j(4\pi/N)} & e^{-j(8\pi/N)} & e^{-j(12\pi/N)} \\ 1 & e^{-j(6\pi/N)} & e^{-j(12\pi/N)} & e^{-j(18\pi/N)} \end{bmatrix}}_{\text{DFT matrix: } F} \begin{bmatrix} 2 \\ 3 \\ -1 \\ 1 \end{bmatrix} = \begin{bmatrix} 5 \\ 3 - j2 \\ -3 \\ 3 + j2 \end{bmatrix}.$$

The above values for the DFT coefficients are the same as the ones obtained in Example 7.1.

Example 7.4

Calculate the inverse DFT of $X[r]$ considered in Example 7.2.

Solution

Arranging the values of the DFT coefficients in the DFT vector x, we obtain

$$X = [5 \quad 3-j2 \quad -3 \quad 3+j2]^T.$$

Using Eq. (7.8), the DFT vector X is given by

$$\begin{bmatrix} x[0] \\ x[1] \\ x[2] \\ x[3] \end{bmatrix} = \frac{1}{4}\begin{bmatrix} 1 & 1 & 1 & 1 \\ 1 & e^{j(2\pi/N)} & e^{j(4\pi/N)} & e^{j(6\pi/N)} \\ 1 & e^{j(4\pi/N)} & e^{j(8\pi/N)} & e^{j(12\pi/N)} \\ 1 & e^{j(6\pi/N)} & e^{j(12\pi/N)} & e^{j(18\pi/N)} \end{bmatrix}\begin{bmatrix} X[0] \\ X[1] \\ X[2] \\ X[3] \end{bmatrix}$$

$$= \frac{1}{4}\begin{bmatrix} 1 & 1 & 1 & 1 \\ 1 & e^{j(2\pi/N)} & e^{j(4\pi/N)} & e^{j(6\pi/N)} \\ 1 & e^{j(4\pi/N)} & e^{j(8\pi/N)} & e^{j(12\pi/N)} \\ 1 & e^{j(6\pi/N)} & e^{j(12\pi/N)} & e^{j(18\pi/N)} \end{bmatrix}\begin{bmatrix} 5 \\ 3-j2 \\ -3 \\ 3+j2 \end{bmatrix} = \frac{1}{4}\begin{bmatrix} 8 \\ 12 \\ -4 \\ 4 \end{bmatrix} = \begin{bmatrix} 2 \\ 3 \\ -1 \\ 1 \end{bmatrix}.$$

The above values for the DT sequence $x[k]$ are the same as the ones obtained in Example 7.2.

7.3 Properties of the DFT

In this section, we present the properties of the M-point DFT. The length of the DT sequence is assumed to be $N \leq M$. For $N < M$, the DT sequence is zero-padded with $M - N$ zero-valued samples. The DFT properties presented below.

7.3.1 Periodicity

The M-point DFT of an aperiodic DT sequence with length N with $M \geq N$ is itself periodic with period M. In other words,
$$X[r] = X[r + pM], \tag{7.9}$$

for $0 \leq r \leq (M - 1)$ with $p \in R^+$.

7.3.2 Orthogonality

The column vectors F_r of the DFT matrix F, defined in Eq. (7.8), form the basis vectors of the DFT and are orthogonal with respect to each other such that

$$F_r^H \cdot F_q = \begin{cases} M & \text{for } r = q \\ 0 & \text{for } r \neq q, \end{cases}$$

where (.) represents the dot product and the superscript H represents the Hermitian operation.

7.3.3 Linearity

If $x_1[k]$ and $x_2[k]$ are two DT sequences with the following M-point DFT pairs:

$$x_1[k] \xleftrightarrow{\text{DFT}} X_1[r] \text{ and } x_2[k] \xleftrightarrow{\text{DFT}} X_2[r],$$

then the linearity property states that

$$a_1 x_1[k] + a_2 x_2[k] \xleftrightarrow{\text{DFT}} a_1 X_1[r] + a_2 X_2[r], \qquad (7.10)$$

for any arbitrary constants a_1 and a_2 which may be complex-valued.

7.3.4 Hermitian symmetry

The M-point DFT $X[r]$ of a real-valued aperiodic sequence $x[k]$ is conjugate-symmetric about $r = m/2$. Mathematically, the Hermitian symmetry implies that

$$X[r] = X^*[M - r], \qquad (7.11)$$

where $X^*[r]$ denotes the complex conjugate of $X[r]$.

In terms of the magnitude and phase spectra of the DFT $X[r]$, the Hermitian symmetry property can be expressed as follows:

$$|X[M-r]| = |X[r]| \text{ and } <X[M-r] = -<X[r] \quad (7.12)$$

implying that the magnitude spectrum is even and that the phase spectrum is odd.

7.3.5 Time shifting

If $x[k] \xleftrightarrow{\text{DFT}} X[r]$, then

$$x[k-k_0] \xleftrightarrow{\text{DFT}} e^{-j2\pi k_0 r/M} X[r] \quad (7.13)$$

for an M-point DFT and any arbitrary integer k_0.

7.3.6 Circular convolution

If $x_1[k]$ and $x_2[k]$ are two DT sequences with the following M-point DFT pairs:

$$x_1[k] \xleftrightarrow{\text{DFT}} X_1[r] \quad \text{and} \quad x_2[k] \xleftrightarrow{\text{DFT}} X_2[r],$$

then the circular convolution property states that

$$x_1[k] \otimes x_2[k] \xleftrightarrow{\text{DFT}} X_1[r]X_2[r] \quad (7.14)$$

and

$$x_1[k]x_2[k] \xleftrightarrow{\text{DFT}} \frac{1}{M}[X_1[r] \otimes X_2[r]], \quad (7.15)$$

where ⊗ denotes the circular convolution operation. Note that the two sequences must have the same length in order to compute the circular convolution.

Example 7.5
Calculated the circular convolution $y[k]$ of the two aperiodic sequences $x[k] = [0, 1, 2, 3]$ and $h[k] = [5, 5, 0, 0]$ defined over $0 \leq k \leq 3$. Recalculate the result of the circular convolution using the DFT convolution property.

Solution
The four-point DFTs of the aperiodic sequences $x[k]$ and $h[k]$ are given by

$$X[r] = [6, -2 + j2, -2, -2 - j2] \text{ and } H[r] = [10, 5 - j5, 0, 5 + j5]$$

for $0 \leq r \leq 3$. Using Eq. (12.27), the four-point DFT of the circular convolution between $x[k]$ and $h[k]$ is given by

$$x_1[k] \otimes x_2[k] \xleftarrow{DFT} [60, j20, 0 - j20].$$

Taking the inverse DFT, we obtain

$$x1[k] \otimes x2[k] = [15, 5, 15, 25],$$

7.3.7 Parseval's theorem

If $x[k] \xleftrightarrow{DFT} X[r]$, then the energy of the aperiodic sequence $x[k]$ of length N can be expressed in terms of its M-point DFT as follows:

$$E_x = \sum_{k=0}^{N-1} |x[k]|^2 = \frac{1}{M} \sum_{k=0}^{M-1} |X[r]|^2.$$

(7.16)

Parseval's theorem shows that the DFT preserves the energy of the signal within a scale factor of M.

7.4 Computational complexity

We now compare the computational complexity of the time-domain and DFT- based implementations of the linear convolution between the time-limited sequences $x_1[k]$ and $x_2[k]$ with lengths K_1 and K_2, respectively. For simplicity, we assume that $x_1[k]$ and $x_2[k]$ are real-valued sequences with lengths K_1 and K_2, respectively.

Time-domain approach This is based on the direct computation of the convolution sum

$$y[k] = x_1[k] * x_1[k] = \sum_{m=-\infty}^{\infty} x_1[m]x_2[k-m],$$

which requires roughly $K_1 \times K_2$ multiplications and $K_1 \times K_2$ additions. The total number of floating point operations (flops) required with the time-domain approach is therefore given by $2K_1 \times K_2$.

DFT-based approach Step 1 of the DFT-based approach computes two $K = K_1 + K_2 - 1$ point DFTs of the DT sequences $x_1[k]$ and $x_2[k]$. The K-point DFT can be implemented using fast Fourier trans- form (FFT) techniques with $0.5K \log_2 K$ complex multiplications and $K \log_2 K$ complex additions. Since each complex multiplication requires four scalar multiplications and two scalar additions, a total of six flops are required per complex multiplication. Each complex addition, on the other hand, requires two scalar additions, leading to two flops per complex addition. Therefore,

- Step 1 of the DFT-based approach requires a total of $2 \times [3K \log_2 K + 2K \log_2 K] = 10K \log_2 K$ flops.
- Step 2 multiplies DFTs for $x_1[k]$ and $x_2[k]$. Each DFT has a length of $K = K_1 + K_2 - 1$ points; therefore, a total of K complex multiplications and $K - 1 \approx K$ complex additions are required. The total number of computations required in Step 1 is therefore given by $8K$ or $8(K_1 + K_2 - 1)$ flops.
- Step 3 computes one inverse DFT based on the FFT implementation requiring $5K \log_2 K$ flops.

The total number of flops required with the DFT-based approach is therefore given by

$$15K \log_2 K + 6K \approx 15K \log_2 K \; flops$$

where $K = K_1 + K_2 - 1$. Assuming $K_1 = K_2$, the DFT-based approach pro- vides a computation saving of $O(\log 2 \; K/K)$ in comparison with the direct computation of the convolution sum in the time domain. Table 7.1 compares the computational complexity of the two approaches for a few selected values of K_1 and K_2. We observe that for sequences with lengths greater than 1000 samples, the DFT-based approach provides significant savings over the direct computation of the circular convolution in the time domain.

Table 7.1. Comparison of the computational complexities of the time-domain versus the DFT-based approaches used to compute the linear convolution

Length K_1 of $x_1[k]$	Length K_2 of $x_2[k]$	Computational complexity, flops	
		Time domain ($2K_1 \times K_2$ flops)	DFT ($15K\log_2 K$ flops)
32	5	320	2792
32	16	1024	3916
32	32	2048	5649
1000	5	10 000	150 171
1000	200	400 000	183 943
1000	1000	2 000 000	328 787

7.5 Fast Fourier Transform (FFT)

The *fast Fourier transform* (FFT) was invented by Cooley and Tukey in 1965. They discovered that the DFT operation could be decomposed into a number of other DFTs of shorter lengths. They then showed that the total number of computations needed for the shorter DFTs was smaller than the number needed for the direct computation. In fact, the number of arithmetic operations (multiplications and additions) for the direct computation of the DFT is approximately equal to N^2, but for the FFT algorithm reduced to approximately $N.\log_2 N$. To take an example, if $N=1024$, the DFT require approximately 10^6 multiplications and additions, whilst the FFT would require $<10^3$, more than 1000 times fewer.

There are several well-known techniques including the radix-2, radix-4, split radix, Winograd, and prime factor algorithms that are used for computing the DFT. These algorithms are referred to as the fast Fourier transform (FFT) algorithms. In this section, we explain the radix-2 decimation-in-time FFT algorithm.

To provide a general frame of reference, let us consider the computational complexity of the direct implementation of the K-point DFT for the time-limited sequence $x[k]$ with length K. Based on its definition,

$$X[r] = \sum_{k=0}^{K-1} x[k] e^{-j(2\pi kr/K)}, \qquad (7.17)$$

K complex multiplications and $K - 1$ complex additions are required to compute a single DFT coefficient. Computation of all K DFT coefficients requires K^2 complex additions and K^2 complex multiplications, where we have assumed K to be large such that $K - 1 \approx K$. In terms of flops, each complex multiplication requires four scalar multi- plications and two scalar additions, and each complex addition requires two scalar additions. Computation of a single DFT coefficient, therefore, requires $8K$ flops. The total number of scalar operations for computing the complete DFT is given by $8K^2$ flops.

We now proceed with the radix-2 FFT decimation-in-time algorithm.

7.5.1 Derivation of the FFT

The decomposition of the DFT is achieved by breaking a signal $x[n]$ down into two shorter, interleaved subsequences. This process is more commonly known as *decimation–in-time* (DIT). Suppose a signal exists with N sample values, where N is an integer power of 2. The signal $x[n]$ is first separated into two subsequences with $N/2$ samples. One subsequence contains the samples with even-numbered values of n in $x[n]$, and the other contains those with odd- numbered values of n. Writing *n(even)* = *2m* and *n(odd)* = *2m*+1. the DFT can be modified to:

$$X[k] = \sum_{m=0}^{\frac{N}{2}-1} x[2m] \cdot W_N^{2mk} + \sum_{m=0}^{\frac{N}{2}-1} x[2m+1] \cdot W_N^{(2m+1)k} \qquad (7.18)$$

From the Argand diagram in Figure 7.2, it can also be shown that $W_N^2 = W_{N/2}^1$ i.e. $W_8^2 = W_4^1$ etc. Hence, the DFT can now be re-written to show that it can be expressed in terms of two N/2-point DFTs.

$$X[k] = \sum_{m=0}^{\frac{N}{2}-1} x[2m] \cdot W_{\frac{N}{2}}^{mk} + W_N^k \sum_{m=0}^{\frac{N}{2}-1} x[2m+1] \cdot W_{\frac{N}{2}}^{mk} \qquad (7.19)$$

$$X[k] = X_1[k] + W_N^k \cdot X_2[k] \qquad (7.20)$$

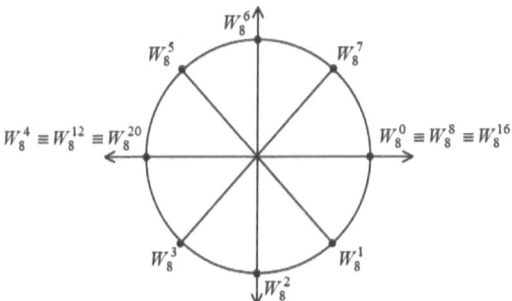

Figure 7.2: Argand diagram illustrating the 8th roots of unity.

$$\begin{aligned} X^{(N)}(k) &= \sum_{n \text{ is even: } n=2m, m=0,1,\cdots,\frac{N}{2}-1} x(n)e^{-j\frac{2\pi k}{N}n} \\ &+ \sum_{n \text{ is odd: } n=2l+1, l=0,1,\cdots,\frac{N}{2}-1} x(n)e^{-j\frac{2\pi k}{N}n} \\ &= \sum_{m=0}^{\frac{N}{2}-1} x(2m)e^{-j\frac{2\pi k}{N}2m} + \sum_{l=0}^{\frac{N}{2}-1} x(2l+1)e^{-j\frac{2\pi k}{N}(2l+1)} \end{aligned} \qquad (7.21)$$

We make the following substitutions

$$x_0(m) = x(2m), \text{ where } m = 0, \cdots, \frac{N}{2} - 1,$$
$$x_1(l) = x(2l+1), \text{ where } l = 0, \cdots, \frac{N}{2} - 1.$$

Rewriting Eq. (1.31), we get

$$X^{(N)}(k) = \sum_{m=0}^{\frac{N}{2}-1} x_0(m) e^{-j\frac{2\pi k}{\frac{N}{2}}m} + e^{-j\frac{2\pi k}{N}} \sum_{l=0}^{\frac{N}{2}-1} x_1(l) e^{-j\frac{2\pi k}{\frac{N}{2}}l}$$
$$= X_0^{(\frac{N}{2})}(k) + e^{-j\frac{2\pi k}{N}} X_1^{(\frac{N}{2})}(k), \qquad (7.22)$$

Where $W_0^{\frac{N}{2}}(k)$ is the $\frac{N}{2}$-point DFT of the even-numbered samples of $x(n)$ and $W_1^{\frac{N}{2}}(k)$ is the is the $\frac{N}{2}$-point DFT of the odd-numbered samples of $x(n)$. Note that both of them are : $\frac{N}{2}$-periodic discrete-time functions.

We have the following algorithm to compute $X^{(N)}(k)$ for $k = 0$, $\cdots,(N-1)$:

1. Compute $W_0^{\frac{N}{2}}(k)$ for k =0, ..., $\frac{N}{2} - 1$

2. Compute $W_1^{\frac{N}{2}}(k)$ for k =0, ..., $\frac{N}{2} - 1$

3. Perform the computation (1.32) with N complex multiplications and N complex additions.

Actually, it is possible to use fewer than N complex multiplications. Let

$$W_N = e^{-j\frac{2\pi}{N}}.$$

Then

$$W_N^{k+\frac{N}{2}} = e^{-j(\frac{2\pi k}{N}+\pi)}$$
$$= -e^{-j\frac{2\pi k}{N}}$$
$$= -W_N^k$$

Therefore,

$$X^{(N)}(k) = X_0^{(\frac{N}{2})}(k) + W_N^k X_1(k) \quad \text{for } k = 0, \cdots, \frac{N}{2} - 1,$$
$$X^{(N)}\left(k + \frac{N}{2}\right) = X_0^{(\frac{N}{2})}(k) - W_N^k X_1(k) \quad \text{for } k = 0, \cdots, \frac{N}{2} - 1,$$

A decimation-in-time FFT algorithm divides up the input data into shorter interleaved subsequences. Take advantage of the symmetry and periodicity of the complex exponential (let $W_N = e^{-j2p/N}$)

- symmetry:
- periodicity:
- Note that two length N/2 DFTs take less computation than one length N DFT: $2(N/2)^2 < N^2$
- Algorithms that exploit computational savings are collectively called *Fast Fourier Transforms*

This type of FFT can be performed using many butterfly operations, as illustrated in Figure 7.3 for the case of $N = 8$. Here it can be seen that the operations are divided up into $log_2 N$.

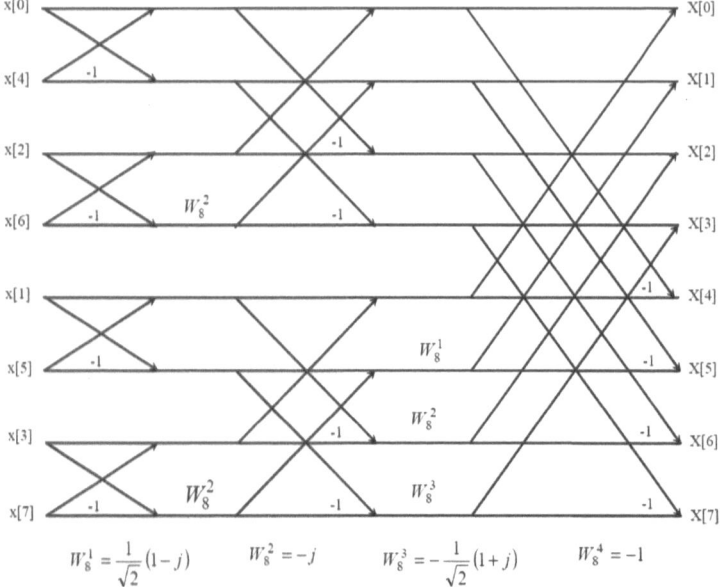

Figure 7.3 Time-decimated radix-2 FFT, N=8.

As illustrated in Figure 7.3. This shows that we do not need to actually perform N complex multiplications, but only $\frac{N}{2}$.

Figure 7.3 illustrates the recursive implementation of the FFT supposing that $N = 2^M$. There is a total of $M = log_2 N$ stages of computation, each requiring $\frac{2}{2} N$ complex operations. Hence, the total computational complexity is $O(N log_2 N)$. We see that the process ends at a 1-point DFT. A 1-point DFT is the sample of the original signal:

$$X(0) = \sum_{n=0}^{0} x(n) e^{-j(\frac{2\pi \cdot 0}{1})n} = x(0).$$

The recursive implementation of the FFT supposing that $N = 2^M$. There is a total of $M = log_2 N$ stages of computation, each requiring $\frac{2}{2}N$ complex operations. Hence, the total computational complexity is $O(N \log N)$.

1. For large N, the FFT is much faster than the direct application of the definition of DFT, which is of complexity $O(N^2)$.
2. The particular implementation of the FFT described above is called *decimation- in-time radix-2 FFT.*
3. The number of operations required by an FFT algorithm can be approximated as $CN \log N$, where C is a constant. There are many variations of FFT aimed at reducing this constant–e.g., if $N = 3^M$, it may be better to use a radix-3 FFT.
4. Note that

$$\left\{\frac{1}{N}\text{DFT}[x^*(n)]\right\}^* = \left\{\frac{1}{N}\sum_{n=0}^{N-1} x^*(n) e^{-j(\frac{2\pi k}{N})n}\right\}^*$$

$$= \frac{1}{N}\sum_{n=0}^{N-1} x(n) e^{j(\frac{2\pi k}{N})n}$$

which is the IDFT of $x(n)$. Thus, the FFT can also be used to compute the IDFT.

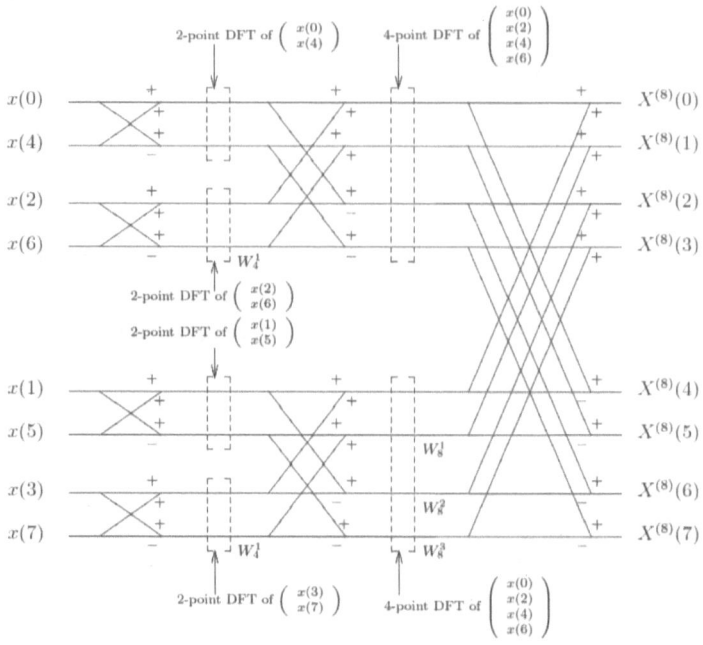

Figure 7.4 The 8-point FFT.

Example 7.6

The 8-point FFT is depicted in Fig. 1.38. The values of the twiddle factors are:

$$W_2 = e^{-j\frac{2\pi}{2}} = -1,$$
$$W_4 = e^{-j\frac{2\pi}{4}} = -j,$$
$$W_8 = e^{-j\frac{2\pi}{8}}.$$

Decimation-in-frequency FFTs are in a sense the exact opposite of the decimation-in-time algorithms; they are simply the consequence of the symmetry of the Fourier transform. A decimation-in-frequency FFT, illustrated in Figure 2.4 for the case o N = 8, uses the opposite approach to the DIT. Here, the output sequence is decimated rather than the input sequence.

Problems

7.1 Determine analytically the DFT of the following time sequences, with length $0 \le k \le (N-1)$:

(i) $x[k] = \begin{cases} 1 & k = 0, 3 \\ 0 & k = 1, 2 \end{cases}$ with length $N = 4$;

(ii) $x[k] = \begin{cases} 1 & k \text{ even} \\ -1 & k \text{ odd} \end{cases}$ with length $N = 8$;

(iii) $x[k] = 0.6^k$ with length $N = 8$;
(iv) $x[k] = u[k] - u[k-8]$ with length $N = 8$;
(v) $x[k] = \cos(\omega_0 k)$ with $\omega_0 \ne 2\pi r/N$.

7.2 Determine the DFT of the time-limited sequences specified in Examples 7.1(i)–(iv) using the matrix-vector approach.

7.3 Determine the time-limited sequence, with length $0 \le k \le (N-1)$, corresponding to the following DFTs $X[r]$, which are defined for the DFT index $0 \le r \le (N-1)$:

(i) $X[r] = [1 + j4, -2 - j3, -2 + j3, 1 - j4]$ with $N = 4$;
(ii) $X[r] = [1, 0, 0, 1]$ with $N = 4$;
(iii) $X[r] = \exp -j(21 r k_o r/N)$, where k_0 is constant

(iv) $X[r] = \begin{cases} 0.5N & r = k_0, N - k_0 \\ 0 & \text{elsewhere} \end{cases}$ where k_0 is a constant;

(v) $X[r] = \begin{cases} k_0 & r = 0 \\ e^{-j(\pi r(N_1-1)/N)} \dfrac{\sin(\pi r k_0/N)}{\sin(\pi r/N)} & r \ne 0 \end{cases}$ where k_0 is a constant;

(vi) $X[r] = \left(\dfrac{r}{N}\right)$ for $0 \le r \le (N-1)$.

7.4 Using MATLAB, compute the DTFT representation based on the FFT algorithm. Plot the frequency characteristics and compare the computed results with the analytical results.

(i) $x[k] = \cos(10\pi k/3)\cos(2\pi k/5)$;
(ii) $x[k] = |\cos(2\pi k/3)|$;
(iii) $x[k] = k$ for $0 \leq k \leq 5$ and $x[k+6] =$;
(iv) $x[k] = \sum_{m=-\infty}^{\infty} \delta(k - 5m)$;

(v) $x[k] = \begin{cases} 1 & 0 \leq k \leq 2 \\ 0.5 & 3 \leq k \leq 5 \\ 0 & 6 \leq k \leq 8 \end{cases}$ and $x[k+9] = x[k]$;

(vi) $x[k] = 2\exp\left(j\frac{5\pi}{3}k + \frac{\pi}{4}\right)$;

(vii) $x[k] = 3\sin\left(\frac{2\pi}{7}k + \frac{\pi}{4}\right)$.

7.5 (a) Using the FFT algorithm in MATLAB, determine the DTFT representation for the following sequences. Plot the magnitude and phase spectra in each case.

(1) $x[k] = k3^{-|k|}$ for all k;
(2) $x[k] = a^k \cos(w_0 k)u[k]$, $|a| < 1$;
(3) $x[k] = -k$;
(4) $x[k] = \sum_{m=-\infty}^{\infty} \delta(k - 5m - 3)$;
(5) $x[k] = \alpha^k \sin(\omega_0 k + \phi)u[k]$, $|\alpha| < 1$;
(6) $x[k] = \dfrac{\sin(\pi k/5)\sin(\pi k/7)}{\pi^2 k^2}$.

(b) Compare the obtained results with the analytical results.

7.6 Using the FFT algorithm in MATLAB, determine the CTFT representation for each of the following CT functions. Plot the frequency characteristics and compare the results with the analytical results.

(i) $x(t) = e^{-2t}u(t)$;

(ii) $x(t) = e^{-4|t|}$;

(iii) $x(t) = t^4 e^{-4t}u(t)$;

(iv) $x(t) = e^{-4t}\cos(10\pi t)u(t)$;

(v) $x(t) = e^{-t^2/2}$;

7.7 Prove the Hermitian property for the DFT.
7.8 Prove the time-shifting property for the DFT.
7.9 Prove the periodic-convolution property for the DFT.
7.10 Prove Parseval's relationship for the DFT.
7.11 Without explicitly determining the DFT $X[r]$ of the time-limited sequence.

$$x[k] = [6\ 8\ -5\ 4\ 16\ 22\ 7\ 8\ 9\ 44\ 2],$$

compute the following functions of the DFT $X[r]$:

(i) $X[0]$;

(ii) $X[10]$;

(iii) $X[6]$;

(iv) $\sum_{r=0}^{10} X[r]$;

(v) $\sum_{r=0}^{10} |X[r]|^2$.

7.12 Without explicitly determining the the time-limited sequence $x[k]$ for the following DFT:

$X[r] = [12\ 8+j4\ -5\ 4+j1\ 16\ 16\ 4-j1\ -5\ 8\ -j4]$,

compute the following functions of the DFT $X[r]$:

(i) $x[0]$;
(ii) $x[9]$;
(iii) $x[6]$;
(iv) $\sum_{r=0}^{9} x[k]$;
(v) $\sum_{r=0}^{9} |x[k]|^2$;

7.13 Draw the flow graph for a 6-point DFT by subdividing into three 2- point DFTs that can be combined to compute $X[r]$. Repeat for the subdivision of two 3-point DFTs. Which one provides more computational savings?

7.14 Draw a flow graph for a 10-point decimation-in-time FFT algorithm using two DFTs of size 5 in the first stage of the flow graph and five DFTs of size 2 in the second stage. Compare the computational complexity of the algorithm with the direct approach based on the definition.

7.15 Assume that $K = 33$. Draw the flow graph for a K-point decimation- in-time FFT algorithm consisting of three stages by using radix-3 as the basic building block. Compare the computational complexity of the rithm with the direct approach based on the definition.

Chapter Eight

Design of Digital Filters

Learning Outcomes of this Chapter

After successful completion of this chapter students will be able to:

1. apply the principles of signal analysis to filtering.
2. describe and learn FIR and IIR filters, their frequency response and characteristics.
3. design and implement FIR and IIIR filters using different methods, and how to test, analyze and refine design.

8.1 Introduction

In signal processing, the function of a filter is to remove unwanted parts of the signal, such as random noise, or to extract useful parts of the signal, such as the components lying within a certain frequency range. The idea of filtering is based on the convolution property of the Fourier Transform. The following block diagram illustrates the basic idea

Figure 8.1 Filtering process

Therefore, a filter is a device that transmits (or rejects) a specific range of frequencies. There are two main kinds of filter, analog and digital. They are quite different in their physical makeup and in how they work.

An analog filter uses analog electronic circuits made up from components such as resistors, capacitors and op amps to produce the required filtering effect. Such filter circuits are widely used in such applications as noise reduction, video signal enhancement, graphic equalizers in hi-fi systems, and many other areas. There are well-established standard techniques for designing an analog filter circuit for a given requirement. At all stages, the signal being filtered is an electrical voltage or current which is the direct analogue of the physical quantity (e.g. a sound or video signal or transducer output) involved.

A digital filter uses a digital processor to perform numerical calculations on sampled values of the signal. The processor may be a general-purpose computer such as a PC, or a specialized DSP (Digital Signal Processor) chip. Table 8.1 shows the types of filters in general.

Table 8.1 shows the types of filters.

Types of filter	Characteristics	Ideal	Practical
Low pass	Passband that extends from $\omega=0$ to $\omega=\omega c$, where ωc is the cutoff frequency		

High pass	Stopband that extends from $\omega=0$ to $\omega=\omega c$, and a passband that extends from $\omega=\omega c$ to ∞	
Band pass	Passband that extends from $\omega=\omega 1$ to $\omega=\omega 2$, and stops all other frequencies	
Band stop	Stop frequencies extending from $\omega=\omega 1$ to $\omega=\omega 2$, and passes other frequencies	

Therefore, the analog input signal must first be sampled and digitized using an ADC (analog to digital converter). The resulting binary numbers, representing successive sampled values of the input signal, are transferred to the processor, which carries out numerical calculations on them. These calculations typically involve multiplying the input values by constants and adding the products together. If necessary, the results of these calculations, which now represent sampled values of the filtered signal, are output through a DAC (digital to analog converter) to convert the signal back to analog form.

Note that in a digital filter, the signal is represented by a sequence of numbers, rather than a voltage or current. There are four basic filter types; *Low-pass, high-pass, band-pass* and *band-stop*. Therefore, A digital filter is a mathematical algorithm implemented in hardware and/or software that operates on a digital input signal to produce a digital output signal for the purpose of achieving a filtering objective.

A *digital filter* uses a digital processor to perform numerical calculations on sampled values of the signal. The processor may be a general-purpose computer such as a PC, or a specialized DSP (Digital Signal Processor) chip. Common filtering objectives are:

a) to improve the quality of a signal
b) to extract information from signals
c) to separate two or more signals previously combined.

The following list gives some of the main advantages of digital over analog filters.

(a) A digital filter is *programmable*, i.e. its operation is determined by a program stored in the processor's memory. This means the digital filter can easily be changed without affecting the circuitry (hardware). An analog filter can only be changed by redesigning the filter circuit.

(b) Digital filters are easily *designed, tested* and *implemented* on a general-purpose computer or workstation.

(c) The characteristics of analog filter circuits (particularly those containing active components) are subject to drift and are dependent on temperature. Digital filters do not suffer from these problems, and so are extremely *stable* with respect both to time and temperature.

(d) Unlike their analogue counterparts, digital filters can handle *low frequency* signals accurately. As the speed of DSP technology continues to increase, digital filters are being applied to high frequency signals in the RF (radio frequency) domain, which in the past was the exclusive preserve of analogue technology.

(e) Digital filters are very much more *versatile* in their ability to process signals in a variety of ways; this includes the ability of some types of digital filter to adapt to changes in the characteristics of the signal.

(f) Fast DSP processors can handle complex combinations of filters in parallel or cascade (series), making the hardware requirements relatively *simple* and *compact* in comparison with the equivalent analog circuitry.

There are also two types of Digital filters; Finite impulse response (FIR) and infinite impulse response (IIR). In general FIR filters can be designed to have exact linear phase and there is also great flexibility in shaping their magnitude response. In addition, FIR filters are inherently more stable and the effects of quantization errors are less severe than IIR filters. Conversely, IIR filters require fewer coefficients than FIR filters for a sharp cut-off frequency response, and analogue filters can only be modelled using IIR filters.

8.1.1. Finite Impulse Response

FIR filters are normally designed to have a linear phase response. Equation 8.1 and Equation 8.2 define the finite difference equation and the z-transfer function for the non-recursive FIR filter respectively. The output of this type of filter is dependent upon the present and previous inputs.

$$y[n] = \sum_{k=0}^{N-1} b_k \cdot x[n-1] \tag{8.1}$$

$$H[z] = \sum_{k=0}^{N-1} b_k \cdot z^{-k} \tag{8.2}$$

8.1.2 Infinite Impulse Response

IIR filter design usually concentrates on the magnitude response and regards the phase response as secondary. Equation 8.3 and Equation 8.4 define the finite difference equation and the z-transfer function for the recursive IIR filter respectively. The output of this filter is dependent upon both previous inputs and outputs.

$$y[n] = \sum_{k=0}^{N} b_k \cdot x[n-1] - \sum_{k=1}^{M} a_k \cdot y[n-1] \quad (8.3)$$

$$H[z] = \frac{\sum_{k=0}^{N} b_k \cdot z^{-k}}{1+\sum_{k=1}^{N} a_k \cdot z^{-k}} \quad (8.4)$$

8.1.3 Filter Specification Requirements

A low-pass filter, as depicted in Figure 8.2 provides a graphical description of the specifications of a normalized low-pass filter. The shaded areas pass low frequencies from zero to a cut off frequency F_p, with approximately unity gain. The frequency range from zero up to F_p is called the *pass band* of the filter. The filter is specifically designed so that any frequencies greater than F_s become attenuated. The frequency range above F_s is referred to as the *stop band* of the filter. Between the pass band and stop band, there is a region called a *transition band* and the exact behaviour of the frequency response in this region is usually of little importance.

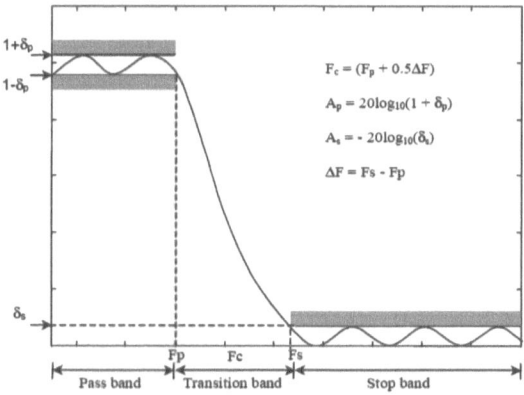

Figure 8.2 A normalized low-pass filter specification.

The parameter δ_s is the tolerance of the magnitude response in the stop-band and the desired magnitude response is always close to zero. The quantity δ_s is known as the *stop-band attenuation* and is sometimes expressed in terms of dBs using the following equation.

Minimum stop-band attenuation, $A_s = -20\log_{10}(\delta_s)$

The parameter δ_p is the tolerance of the magnitude response in the pass-band and the desired magnitude response is close to unity. The quantity δ_p is called the *pass-band ripple* and can also be expressed in terms of dBs using the following equation.

Pass-band ripple, $A_p = 20\log_{10}(1 + \delta_p)$

8.2 FIR Digital Filters

FIR filters have characteristics that make them useful in many applications.

1. FIR filters can achieve an exactly linear phase frequency response.
2. FIR filters cannot be unstable.

3. FIR filters are generally less sensitive to coefficient round-off and finite-precision arithmetic than IIR filters.
4. FIR filters design methods are generally linear.
5. FIR filters can be efficiently realized on general or special-purpose hardware.

However, frequency responses that need a rapid transition between bands and do not require linear phase are often more efficiently realized with IIR filters.

The transfer function of an FIR filter, in particular, is given by

$$H(z^{-1}) = b_0 + b(1)z^{-1} + b(2)z^{-2} + \cdots + b(M)z^{-M} \qquad (8.5)$$

and the difference equation describing this FIR filter is given by

$$y[n] = \sum_{k=0}^{M} b(k).x[n-1] \qquad (8.6)$$

$$= b(0)x9n \mp b(1)x(n-1) + \ldots + b(M)x(n-M) \quad (8.7)$$

In general FIR filters can be designed to have exact linear phase and there is also great flexibility in shaping their magnitude response. In addition, FIR filters are inherently more stable, and the effects of quantization errors are less severe than IIR filters. Conversely, IIR filters require fewer coefficients than FIR filters for a sharp cut-off frequency response, and analogue filters can *only* be modelled using IIR filters. In this section, the properties of the FIR filters and their design will be discussed. When the input function *x(n)* is the unit sample function *δ(n)*, the output *y(n)* can be obtained by applying the recursive algorithm on (8.6). We get the output *y(n)* due to the unit sample input *δ(n)* to be exactly the values *b(0), b(1), b(2), b(3), ..., b(M)*. The output due to the unit sample function *δ(n)* is the unit sample response or the unit impulse response denoted by $h(n)$. So the samples of the unit impulse response $h(n)$ $b(n)$, which means that the unit impulse response $h(n)$ of the discrete-time system described by the difference equation (8.6) is finite in length. That is why the system is called the *finite impulse response filter* or the FIR filter. It has also been known by other names such as the transversal filter, nonrecursive filter, moving-average filter, and tapped delay filter. Since $h(n)$ $b(n)$ in the case of an FIR filter, we

$$H(z^{-1}) = \sum_{k=0}^{M} h(k)z^{-k} = h(0) + h(1)z^{-1} + h(2)z^{-2} + \cdots + h(M)z^{-(M)}$$
(8.8)

It is to be remembered that we choose the order of the FIR filter or degree of the polynomial;

$$H(z^{-1}) = \sum_{n=0}^{N} h(n)z^{-n}$$

As N, and the length of the filter equal to the number of coefficients in (8.8) is N 1. If we are given $H(z^{-1}) = 0.3z^{-4} + 0.1z^{-5} + 0.5z^{-6}$, its order is 6, although only three terms are present and the correct number of coefficients equal to the length of the filter is 7, because $h(0)$, $h(1)$, $h(2)$, $h(3) = 0$. It becomes necessary to point out the notation used in this chapter, because in some textbooks, we may find $H(z^{-1})$ $^{N-1}$ $h(n)z^{-n}$ representing the transfer function of an FIR filter, in which case the length of the filter is denoted by N and the degree or order of the polynomial is $(N-1)$.

Consider the *ideal* low-pass filter frequency response, as illustrated in Figure 8.3 below, with a normalized angular cut-off frequency Ω_c. Usually, the subscript D is used to distinguish between the ideal and actual, impulse and frequency responses of a filter.

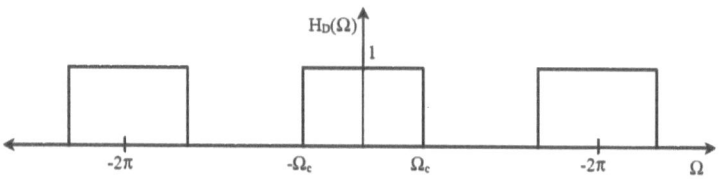

Figure 8.3 Ideal low-pass filter frequency response

The impulse response of an ideal low-pass filter $h_D[n]$ is found by substituting $H_D[\Omega] = 1$ and integrating between the limits of the cut-off frequencies $[-\Omega_c, \Omega_c]$.

$$h_D[n] = \frac{1}{2\pi}\int_{-\Omega_c}^{\Omega_c} 1 \cdot e^{jn\Omega} d\Omega = \frac{1}{2\pi}\left[\frac{e^{jn\Omega}}{jn}\right]_{-\Omega_c}^{\Omega_c} = \frac{1}{2\pi}\left[\frac{e^{jn\Omega_c}}{jn} - \frac{e^{-jn\Omega_c}}{jn}\right] = \frac{1}{2\pi} \cdot \frac{2j\sin(n\Omega_c)}{jn}$$

Multiplying both the numerator and denominator by Ω_c, the Equation above becomes:

$$h_D[n] = \frac{\Omega_c}{\pi} \cdot \frac{\sin(n\Omega_c)}{(n\Omega_c)}$$

The impulse response of an ideal low-pass filter can also be re-written by replacing $\Omega_c = 2\pi F_c$ in t above, to obtain it in terms of the normalized cut-off frequency F_c. This is illustrated in Equation 8.9 below.

$$h_D[n] = 2F_c \cdot \frac{\sin(n\Omega_c)}{(n\Omega_c)} \tag{8.9}$$

In Chapter 7, we found that the Fourier transform of a rectangular window is a sinc function, which is same for the impulse response of a low-pass filter, as illustrated in Figure 8.4 below.

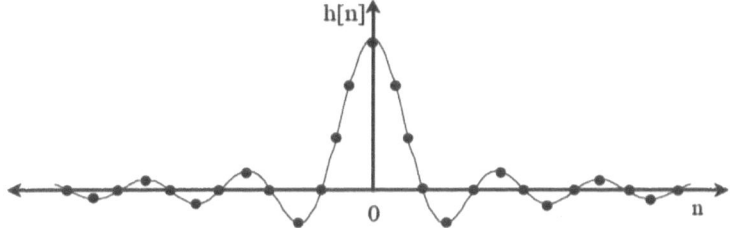

Figure 8.4 Impulse response of an ideal low-pass filter.

Example 8.1.
Given the following FIR filter:

$$y(n) = 0.1x(n) + 0.25\,x(n-1) + 0.2x(n-2)$$

Determine the transfer function, filter length, nonzero coefficients, and impulse response.

Solution:
Applying z-transform on both sides of the difference equation yields

$$Y(z) = 0.1X(z) + 0.25X(z)z^{-1} + 0.2X(z)z^{-2}.$$

Then the transfer function is found to be

$$H(z) = \frac{Y(z)}{X(z)} = 0.1 + 0.25z^{-1} + 0.2z^{-2}.$$

The filter length is K +1 =3 and the identified coefficients are

$$b_0 = 0.1, \quad b_1 = 0.25 \quad and \quad b_2 = 0.2$$

Taking the inverse z-transform of the transfer function, we have

$$h(n) = 0.1\delta(n) + 0.25\delta(n-1) + 0.2\delta(n-2).$$

This FIR filter impulse response has only three terms.

The foregoing example is to help us understand the FIR filter format. We can conclude that:

- The transfer function has a constant term, all the other terms each have a negative power of z, and all the poles are at the origin on the z-plane. Hence, the stability of filter is guaranteed. Its impulse response has only a finite number of terms.
- The FIR filter operations involve only multiplying the filter inputs by their corresponding coefficients and accumulating them; the implementation of this filter type in real time is straightforward.

From the FIR filter format, the design objective can be to obtain the FIR filter b_i coefficients such that the magnitude frequency response of the FIR filter $H(z)$ will approximate the desired magnitude frequency response, such as that of a lowpass, highpass, bandpass, or bandstop filter. The following sections will introduce design methods to calculate the FIR filter coefficients.

Digital FIR filters have many favorable properties, which is why they are extremely popular in digital signal processing. One of these properties is that they may exhibit linear phase, which means that signals in the passband will suffer no dispersion. Dispersion occurs when different frequency components of a signal have a different delay through a system.

The design of a digital filter is carried out in three steps:

- Specifications: they are determined by the applications

- *Approximations/coefficients calculation:* once the specification are defined, we use various concepts and mathematics that we studied so far to come up with a filter description that approximates the given set of specifications (H(z)).

Realization/Implementation: The product of the above step is a filter description in the form of either a difference equation, or a system function H(z), or an impulse response h(n). From this description we implement the filter in hardware or through software on a computer.

There are many methods to design FIR filters such *impulse response truncation, windowing* technique and using *frequency sampling.*

8.2.1 Design of FIR Digital Filters using Impulse Response Truncation (IRT)

With reference to Figure 8.4, although $h[n]$ decays to either side of $n = 0$ it theoretically continues for ever in both directions. This reflects a general antithesis between band limitation and time limitation; since we have chosen a frequency response with a sharp cut-off (or brick wall response), then the time-domain response continues forever. To realize such a filter the impulse response is *truncated* in some way or other. One approach is to ignore the small sample values at the ends and shift $h[n]$ to begin at $n = 0$, giving a *causal filter* as depicted in Figure 8.5. In general, the more samples we include of $h[n]$ the closer we get to the desired form of $H_D[\Omega]$, but the less economic the filter becomes due to the relative number of computations. In practice we must settle for an approximation to the ideal frequency response. It is usually customary to truncate the impulse response to N = (2M + 1) terms. In Figure 8.5, the impulse response of the ideal low-pass filter is truncated to M = 9 samples and is delayed by M samples.

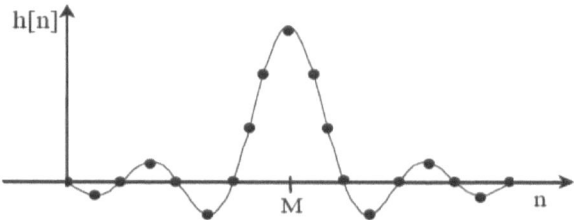

Figure 8.5 Truncated impulse response to 2M+1 samples, with a delay of M samples.

The z-transfer function of the filter now becomes:

$$H(z) = \sum_{n=-M}^{M} h[n] \cdot z^{-(n+M)} \tag{8.10}$$

The ideal impulse responses for a low-pass, high-pass, band-pass and band-stop filters are depicted in Table 8.2 below.

Table 8.2 Ideal impulse responses for various FIR filter types

Filter type	$h_D[n], n \neq 0$	$h_D[n], n = 1$
Low-pass	$2F_c \dfrac{\sin(n\Omega_c)}{n\Omega_c}$	$2F_c$
High-pass	$1 - 2F_c \dfrac{\sin(n\Omega_c)}{n\Omega_c}$	$1 - 2F_c$
Band-pass	$2F_2 \dfrac{\sin(n\Omega_2)}{n\Omega_2} - 2F_1 \dfrac{\sin(n\Omega_1)}{n\Omega_1}$	$2F_2 - 2F_1$
Band-stop	$1 - \left[2F_2 \dfrac{\sin(n\Omega_1)}{n\Omega_1} - 2F_1 \dfrac{\sin(n\Omega_2)}{n\Omega_2} \right]$	$1 - [2F_2 - 2F_1]$

Summary of FIR Filter Design Using The IRT Method

1. Choose the ideal frequency response $H_D(\Omega)$, depending on the type of filter (e.g. low-pass, high-pass etc), from Table 8.2.
2. Calculate the impulse response of the ideal filter $h_D[n]$, using the inverse Fourier transform.
3. Finally, truncate the ideal impulse response to $N = (2M + 1)$ terms.

8.2.2 Design of FIR filters using windowing technique.

Windowing method is applied to the impulse response of a FIR filter to attenuate the Gibb's oscillations. It is developed to remedy the undesirable Gibbs oscillations in the passband and stopband of the designed FIR filter. It is seeking a window function, which is symmetrical and can gradually weight the designed FIR coefficients down to zeros at both ends for the range of $-M \leq n \leq M$. Applying the window sequence to the filter coefficients gives. Applying the window sequence to the filter coefficients gives:

$$h_w(n) = h(n).x(n),$$

where $w(n)$ designates the window function. Common window functions used in the FIR filter design as shown in Table.8.3

Table 8.3 Common window functions (normalized) used in the FIR filter design.

Window Type	Window Function $w(n)$ $-M \leq n \leq M$	Transition width ΔF in (Hz), (normalized)	Passband Ripple (dB)	Stopband Attenuation (dB)

Rectangular	1	$\Delta f = 0.9/N$	0.7416	21
Hanning	$0.5 + 0.5\cos(\frac{\pi n}{M})$	$\Delta f = 3.1/N$	0.0546	44
Hamming	$0.54 + 0.46\cos(\frac{\pi n}{M})$	$\Delta f = 3.3/N$	0.0194	53
Blackman	$0.42 + 0.5\cos(\frac{\pi n}{M}) + 0.8\cos(\frac{\pi n}{M})$	$\Delta f = 5.5/N$	0.0017	74
Kaiser $\beta=4.54$ $\beta = 5.65$ $\beta = 6.76$ $\beta = 8.96$	$I_0\left[\beta\sqrt{1-\left(\frac{\|2n-N+1\|}{N-1}\right)^2}\right] - I_0(\beta)$ where $I_0(x) = \sum_{k=0}^{\infty}\left(\frac{x^k}{2^k k!}\right)^2$	$\Delta f = 2.93/N$ $\Delta f = 3.63/N$ $\Delta f = 4.32/N$ $\Delta f = 5.71/N$	0.0274 0.00867 0.00275 0.000275	50 60 70 90

In addition, there is another popular window function, called the Kaiser window (its detailed information can be found in Oppenheim, Schafer, and Buck [1999]). As we expected, the rectangular window function has a constant value of 1 within the window, hence does only truncation. As a comparison, shapes of the other window functions. Figure 8,6 shows the graphical representation and its frequency spectrum of the mentioned common windows.

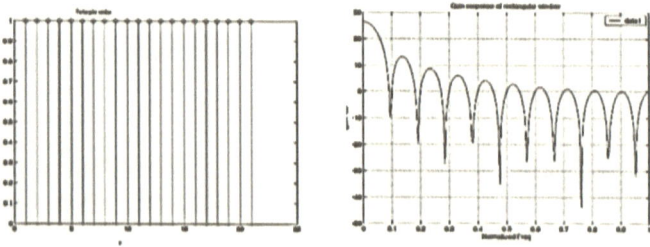

(a) Rectangular window

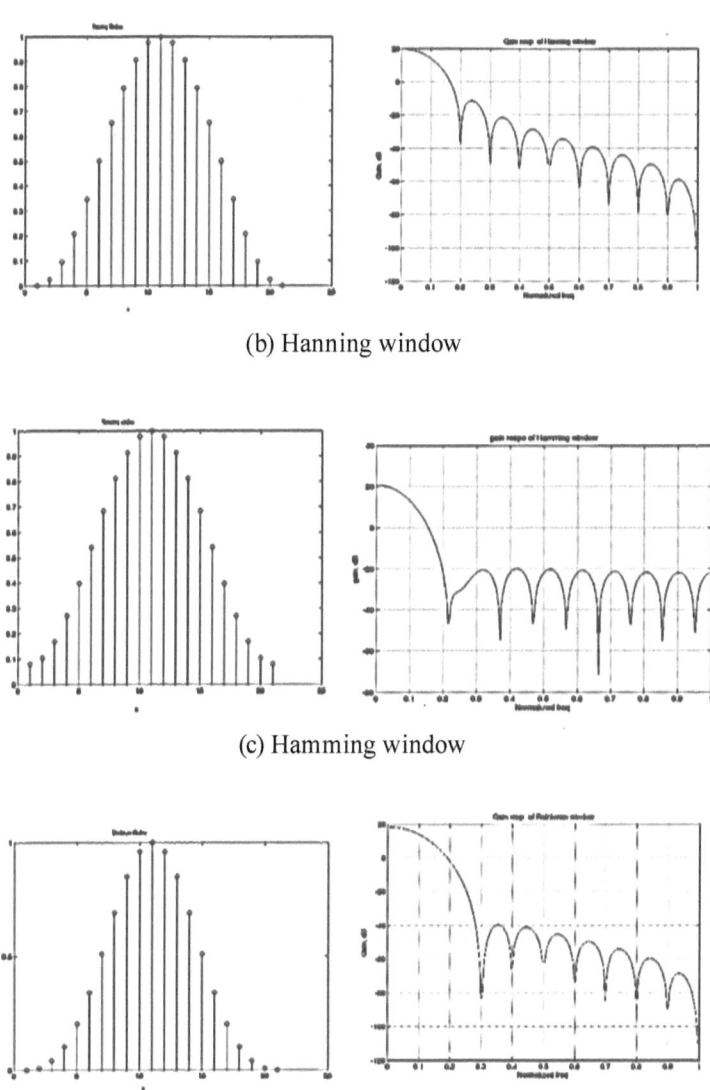

(b) Hanning window

(c) Hamming window

(c) Blackman Windows

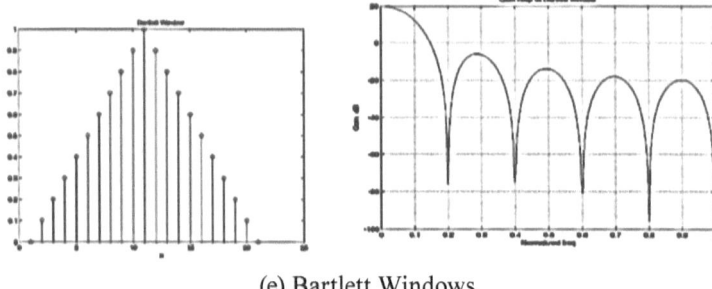

(e) Bartlett Windows
Figure 8.6 Common window and its frequency spectrum

When designing a FIR filter using the window method, it is customary to start by defining the ideal frequency response $H_D[\Omega]$ of the filter. Once the ideal frequency response has been obtained, it is then necessary to obtain the ideal impulse response $h_D[n]$ from the inverse Fourier transform of the frequency response. The next step in the design process is to select an appropriate window function based on the specification of the pass- band δ_p and stop-band δ_s tolerances of the filter. The tolerances of a FIR filter ultimately depend upon the type of window function used for the windowing but in the design process we always start by assuming that $\delta_s = \delta_p$. In Table 8.3, it was illustrating various window functions and their corresponding properties to enable in the design of a filter. Once the appropriate window function has been determined by the designer based on the filter specification, then the required number of filter coefficients N can then be calculated. The final procedure of the design process is to determine the actual filter coefficients $h[n]$ by multiplying the coefficients of the window function with the corresponding ideal impulse response. This is mathematically shown in the following equation :

$$h[n] = w[n] \cdot h_D[n]$$

Steps to design FIR Filter using Window Method

1. Specify the ideal or desired frequency response of the filter $H_D[\Omega]$.
2. Obtain the impulse response $h_D[n]$ by evaluating the inverse Fourier transform.
3. Select an appropriate window function $w[n]$, that satisfies pass-band and attenuation specifications, and determine the number of coefficients required from the relationship between N and Δf.
4. Determine the values of the window function and calculate the *actual* FIR filter coefficients by multiplying the impulse response with the window function.

Example 8.2

a. Design a 3-tap FIR lowpass filter with a cutoff frequency of 800 Hz and a sampling rate of 8,000 Hz using the Hamming window function.
b. Determine the transfer function and difference equation of the designed FIR system.

Solution:

a. The normalized cutoff frequency is calculated as
$$\Omega_c = 2\pi f_c T_s = 2\frac{\pi \times 800}{8000} = 0.2\pi \text{ radians}$$

Since $2M + 1 = 3$ in this case, using the equation in Table 7.1 results in

$$h(0) = \frac{\Omega_c}{\pi} \quad \text{for } n = 0$$

$$h(n) = \frac{\sin(\Omega_c n)}{n\pi} = \frac{\sin(0.2\pi n)}{n\pi}, \quad \text{for } n \neq 1.$$

The computed filter coefficients via the previous expression are listed as:

$$h(0) = \frac{0.2\pi}{\pi} = 0.2$$

$$h(1) = \frac{\sin[0.2\pi \times 1]}{1 \times \pi} = 0.1871.$$

Applying the Hamming window function defined in Table 8.3, we have

$$w_{ham}(0) = 0.54 + 0.46 \cos\left(\frac{0\pi}{1}\right) = 1$$

$$w_{ham}(1) = 0.54 + 0.46 \cos\left(\frac{1 \times \pi}{1}\right) = 0.08.$$

Using the symmetry of the window function gives

$$w_{ham}(-1) = w_{ham}(1) = 0.08.$$

The windowed impulse response is calculated as

$$h_w(0) = h(0)w_{ham}(0) = 0.2 \times 1 = 0.2$$
$$h_w(1) = h(1)w_{ham}(1) = 0.1871 \times 0.08 = 0.01497$$
$$h_w(-1) = h(-1)w_{ham}(-1) = 0.1871 \times 0.08 = 0.01497.$$

Thus, delaying hw(n) by M = 1 sample gives

$$b0 = b2 = 0.01496 \quad and \quad b1 = 0.2$$

b. The transfer function is achieved as

$$H(z) = 0.01496 + 0.2z^{-1} + 0.01496z^{-2}$$

Using the transfer function technique, we have

$$\frac{Y(z)}{X(z)} = H(z) = 0.01497 + 0.2z^{-1} + 0.01497z^{-2}.$$

Multiplying $X(z)$ leads to

$$Y(z) = 0.01497X(z) + 0.2z^{-1}X(z) + 0.01497z^{-2}X(z).$$

Applying the inverse z-transform on both sides, the difference equation is yielded as

$$y(n) = 0.01497x(n) + 0.2x(n-1) + 0.01497x(n-2).$$

Example 8.3
A lowpass FIR filter has the following specifications:
  Passband = 0 - 1,850 Hz
  Stopband = 2,150 - 4,000 Hz Stopband attenuation = 20 dB
  Passband ripple = 1 dB Sampling rate = 8,000 Hz

Determine the FIR filter length and the cutoff frequency to be used in the design equation.

Solution
The normalized transition band as defined $\Delta f = |f_{stop} - f_{pass}|/f_s$ and Table 8.3 is given by;

$$\Delta f = |2150 - 1850|/8000 = 0.0375.$$

Again, based on Table 8.3, selecting the rectangular window will result in a passband ripple of 0.74 dB and a stopband attenuation of 21 dB. Thus, this window selection would satisfy the design requirement for the passband ripple of 1 dB and stopband attenuation of 20 dB. Next, we determine the length of the filter as

$$N = 0.9/\Delta f = 0.9/0.0375 = 24.$$

We choose the odd number $N = 25$. The cutoff frequency is determined by $(1850 + 2150)/2 = 2000$ Hz. Such a filter has coefficients are listed in Table 8.4, and its frequency responses can be found in Figure 8.7 (dashed lines).

Table 8.4 FIR filter coefficients in (rectangular and Hamming windows).

B: FIR Filter Coefficients (rectangular window)	Bham: FIR Filter Coefficients (Hamming window)
$b_0 = b_{24} = 0.000000$	$b_0 = b_{24} = 0.000000$
$b_1 = b_{23} = -0.028937$	$b_1 = b_{23} = -0.002769$
$b_2 = b_{22} = 0.000000$	$b_2 = b_{22} = 0.000000$
$b_3 = b_{21} = 0.035368$	$b_3 = b_{21} = 0.007595$
$b_4 = b_{20} = 0.000000$	$b_4 = b_{20} = 0.000000$
$b_5 = b_{19} = -0.045473$	$b_5 = b_{19} = -0.019142$
$b_6 = b_{18} = 0.000000$	$b_6 = b_{18} = 0.000000$
$b_7 = b_{17} = 0.063662$	$b_7 = b_{17} = 0.041957$
$b_8 = b_{16} = 0.000000$	$b_8 = b_{16} = 0.000000$
$b_9 = b_{15} = -0.106103$	$b_9 = b_{15} = -0.091808$
$b_{10} = b_{14} = 0.000000$	$b_{10} = b_{14} = 0.000000$
$b_{11} = b_{13} = 0.318310$	$b_{11} = b_{13} = 0.313321$
$b_{12} = 0.500000$	$b_{12} = 0.500000$

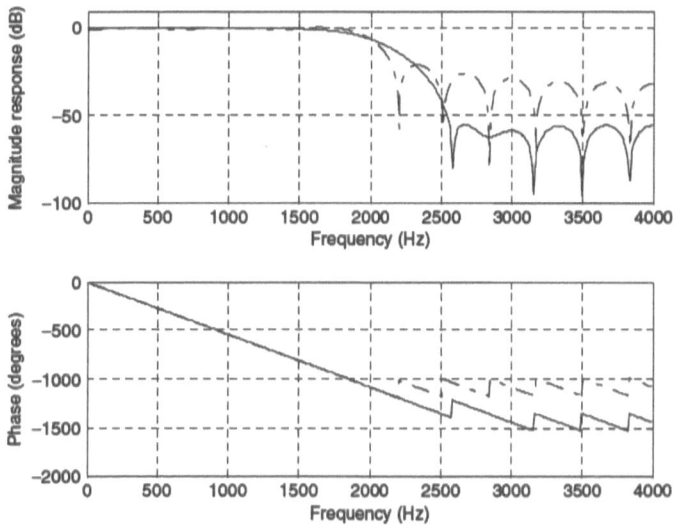

Figure 8.7 Frequency responses using the rectangular and Hamming windows.

8.2.3 Design of FIR filters by frequency sampling

In addition to methods of Fourier transform design and windowing discussed in the previous section, *frequency sampling* is another alternative. The key feature of frequency sampling is that the filter coefficients can be calculated based on the specified magnitudes of the desired filter frequency response uniformly in frequency domain. Hence, it has design flexibility.

To begin with development, we let $h(n)$, for $n = 0, 1, \ldots, N - 1$, be the causal impulse response (FIR filter coefficients) that approximates the FIR filter, and we let $H(k)$, for $k = 0, 1, \ldots, N - 1$, represent the corresponding discrete Fourier transform (DFT) coefficients. We obtain $H(k)$ by sampling the desired frequency filter response $H(k) = H(e^{j\Omega})$ at equally spaced instants in frequency domain, as shown in Figure 8.8. Then, according to the definition of the inverse DFT (IDFT), we can calculate the FIR coefficients:

$$h(n) = \frac{1}{N} \sum_{k=0}^{N-1} H(k) W_N^{-kn}, \text{ for } n = 0, 1, \ldots, N-1,$$

where

$$W_N = e^{-j\frac{2\pi}{N}} = \cos\left(\frac{2\pi}{N}\right) - j\sin\left(\frac{2\pi}{N}\right). \tag{8.11}$$

We assume that the FIR filter has linear phase and the number of taps $N = 2M + 1$. Equation (8.11) can be significantly simplified as

$$h(n) = \frac{1}{2M+1}\left\{H_0 + 2\sum_{k=1}^{M} H_k \cos\left(\frac{2\pi k(n-M)}{2M+1}\right)\right\},$$

for $n = 0, 1, \ldots, 2M$, \hfill (8.12)

where H_k for $k = 0, 1, \ldots, 2M$, represents the magnitude values specifying the desired filter frequency response sampled at $\Omega_k = \frac{2\pi k}{(2M+1)}$. The derivation is

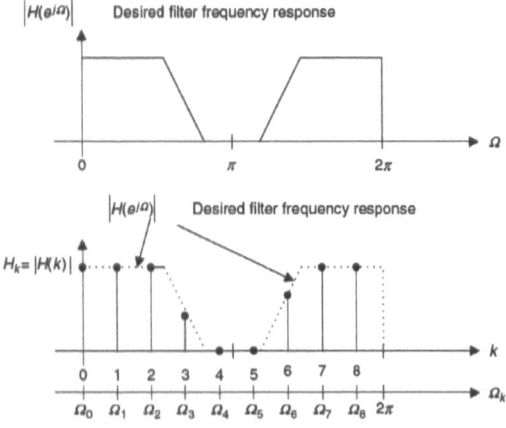

Figure 8.8 Desired filter frequency response and sampled frequency response.

The design procedure is therefore simply summarized as follows:

1. Given the filter length of $2M + 1$, specify the magnitude frequency response for the normalized frequency range from 0 to π.

$$H_k \text{ at } \Omega_k = \frac{2\pi k}{(2M+1)} \quad \text{for } k = 0, 1, \ldots, M. \tag{8.13}$$

2. Calculate FIR filter coefficients:

$$h(n) = \frac{1}{2M+1}\left\{ H_0 + 2\sum_{k=1}^{M} H_k \cos\left(\frac{2\pi k(n-M)}{2M+1}\right)\right\}$$
for $n = 0, 1, \ldots, M$. \hfill (8.14)

3. Use the symmetry (linear phase requirement) to determine the rest of the coefficients:

$$h(n) = h(2M - n) \quad \text{for } n = M + 1, \ldots, 2M \tag{8.15}$$

Example 8.4

Design a linear phase lowpass FIR filter with 7 taps and a cutoff frequency of $\Omega_c = 0.3\pi$ radian using the frequency sampling method.

Solution:

Since $N = 2M + 1 = 7$ and $M = 3$, the sampled frequencies are given by

$$\Omega_k = \frac{2\pi}{7}k \text{ radians}, k = 0, 1, 2, 3.$$

Next we specify the magnitude values H_k at the specified frequencies as follows:

for $\Omega_0 = 0$ radians, $H_0 = 1.0$

for $\Omega_1 = \frac{2}{7}\pi$ radians, $H_1 = 1.0$

for $\Omega_2 = \frac{4}{7}\pi$ radians, $H_2 = 0.0$

for $\Omega_3 = \frac{6}{7}\pi$ radians, $H_3 = 0.0..$

Figure 8.9 shows the specifications

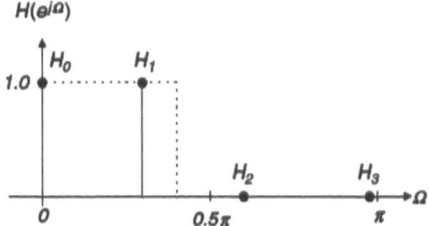

Figure 8.9 Sampled values of the frequency response in Example 8.4.

Using Equation (8.14), we achieve

$$h(n) = \frac{1}{7}\left\{1 + 2\sum_{k=1}^{3} H_k \cos[2\pi k(n-3)/7]\right\}, n = 0, 1, \ldots, 3.$$

$$= \frac{1}{7}\{1 + 2\cos[2\pi(n-3)/7]\}$$

Thus, computing the FIR filter coefficients yields

$$h(0) = \frac{1}{7}\{1 + 2\cos(-6\pi/7)\} = -0.11456$$

$$h(1) = \frac{1}{7}\{1 + 2\cos(-4\pi/7)\} = 0.07928$$

$$h(2) = \frac{1}{7}\{1 + 2\cos(-2\pi/7)\} = 0.32100$$

$$h(3) = \frac{1}{7}\{1 + 2\cos(-0 \times \pi/7)\} = 0.42857.$$

By the symmetry, we obtain the rest of the coefficients as follows:

$$h(4) = h(2) = 0:32100$$
$$h(5) = h(1) = 0:07928$$
$$h(6) = h(0) = -0:11456$$

8.3 Design of IIR Filters

IIR filter design primarily concentrates on the magnitude response of the filter and regards the phase response as secondary. It is a system where the output of the system not only depend on the input signals but the past values of the output signals. The IIR filter characteristic are:
- The system only has both zeros and poles.
- The system has feedback.
- The stability of the system depends on its poles.

The most common design method for digital IIR filters is based on designing an analogue IIR filter and then converting it to an equivalent digital filter. There are many classes of analogue low-pass filter, such as the Butterworth, Chebyshev and Elliptic filters. The classes differ in their nature of their magnitude and phase responses. The design of analogue filters other than low-pass is based on *frequency transformations*, which produce an equivalent high-pass, band-pass, or band-stop filter from a prototype low-pass filter of the same class. The analogue IIR filter is then converted into a similar digital filter using a relevant transformation method. There are three main methods of transformation, the *impulse invariant* method, the *backward difference* method, and the *bilinear z-transform*.

5.3.1 IIR Filter Basics

A *recursive filter* involves *feedback*. In other words, the output values are calculated using one or more of the previous outputs, as well as inputs. In most cases a recursive filter has an impulse response which theoretically continues forever. It is therefore referred to as an infinite impulse response (IIR) filter. Assuming the filter is causal, so that the impulse response $h[n] = 0$ for $n < 0$, it follows that $h[n]$ cannot be symmetrical in form. Therefore, an IIR filter cannot display pure linear-phase characteristics like its adversary, the FIR filter. An IIR filter is described using the difference equation,

$$y(n) = b_0 x(n) + b_1 x(n-1) + \cdots + b_M x(n-M) \\ - a_1 y(n-1) - \cdots - a_N y(n-N). \quad (8.16)$$

also gives the IIR filter transfer function as

$$H(z) = \frac{Y(z)}{X(z)} = \frac{b_0 + b_1 z^{-1} + \cdots + b_M z^{-M}}{1 + a_1 z^{-1} + \cdots + a_N z^{-N}}, \quad (8.17)$$

where b_i and a_i are the $(m-1)$ numerator and N denominator coefficients, respectively. $Y(z)$ and $X(z)$ are the z-transform functions of the filter input $x(n)$ and filter output $y(n)$. To become familiar with the form of the IIR filter, let us look at the following example.

Example 8.5
Given the following IIR filter:

$$y(n) = 0.2x(n) + 0.4x(n-1) + 0.5y(n-1)$$

Determine the transfer function, nonzero coefficients, and impulse response.

Solution:

Applying the z-transform and solving for a ratio of the z-transform output over input, we have

$$H(z) = \frac{Y(z)}{X(z)} = \frac{0.2 + 0.4z^{-1}}{1 - 0.5z^{-1}}.$$

We also identify the nonzero numerator coefficients and denominator coefficient as

$$b_0 = 0.2, \ b_1 = 0.4, \text{ and } a_1 = -0.5:$$

To solve the impulse response, we rewrite the transfer function as

$$H(z) = \frac{0.2}{1-0.5z^{-1}} + \frac{0.4z^{-1}}{1-0.5z^{-1}}$$

Using the inverse z-transform and shift theorem, we obtain the impulse response as

$$h(n) = 0.2(0.5)^n u(n) + 0.4(0.5)^{n-1} u(n-1)$$

The obtained impulse response has an infinite number of terms, where the first several terms are calculated as

$$h(0) = 0.2, \ h(1) = 0.7, \ h(2) = 0.25, \ \ldots$$

At this point, we can make the following remarks:

1. The IIR filter output $y(n)$ depends not only on the current input $x(n)$ and past inputs $x(n-1)$, ..., but also on the past output(s) $y(n-1)$, ... (recursive terms). Its transfer function is a ratio of the numerator polynomial over the

denominator polynomial, and its impulse response has an infinite number of terms.
2. Since the transfer function has the denominator polynomial, the pole(s) of a designed IIR filter must be inside the unit circle on the z-plane to ensure its stability.
3. Compared with the finite impulse response (FIR) filter, the IIR filter offers a much smaller filter size. Hence, the filter operation requires a fewer number of computations, but the linear phase is not easily obtained. The IIR filter is thus preferred when a small filter size is called for but the application does not require a linear phase.

The objective of IIR filter design is to determine the filter numerator and denominator coefficients to satisfy filter specifications such as passband gain and stopband attenuation, as well as cutoff frequency/frequencies for the low- pass, highpass, bandpass, and bandstop filters.

We first focus on the bilinear transformation (BLT) design method. Then we introduce other design methods such as the impulse invariant design and the pole-zero placement design.

There are few methods to design IIR filters such as *Bilinear transformation method*;

8.3.2 Bilinear transformation method

One of the most effective and widely used techniques for converting an *analogue* filter into a *digital* equivalent is by means of the bilinear z-transform. Figure 8.10 illustrates a flow chart of the BLT design used in this book. The deign procedure includes the following steps: (1) transforming digital filter specifications into analog filter specifications, (2) performing analog filter design, and (3) applying bilinear transformation and verifying its frequency response.

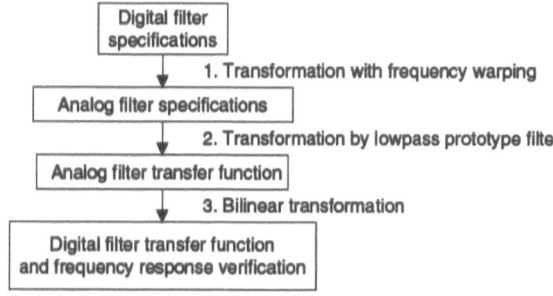

Figure 8.10 General procedure for IIR filter design using bilinear transformation

8.3.3 Analog Filter using lowpass prototype Transformation

Before we begin to develop the BLT design, let us review analog filter design using *lowpass prototype transformation*. This method converts the analog low-pass filter with a cutoff frequency of 1 radian per second, called the lowpass prototype, into practical analog lowpass, highpass, bandpass, and bandstop filters with their frequency specifications. Letting $H_p(s)$ be a transfer function of the lowpass prototype, the transformation of the lowpass prototype into a lowpass filter is given in Figure 8.11.

As shown in Figure 8.11, $H_{LP}(s)$ designates the analog lowpass filter with a cutoff frequency of v_c radians/second. The lowpass-prototype to lowpass-filter transformation substitutes s in the lowpass prototype function $H_p(s)$ with $s=v_c$, where v is the normalized frequency of the lowpass prototype and v_c is the cutoff frequency of the lowpass filter to be designed. Let us consider the following first-order lowpass prototype:

$$H_P(s) = \frac{1}{s+1}. \tag{8.18}$$

Its frequency response is obtained by substituting $s = j\nu$ into Equation (8.18), that is,

$$|H_P(j\nu)| = \frac{1}{\sqrt{1+\nu^2}}.$$

with the magnitude gain given in Equation (8.2):

$$|H_P(j\nu)| = \frac{1}{\sqrt{1+\nu^2}}. \tag{8.19}$$

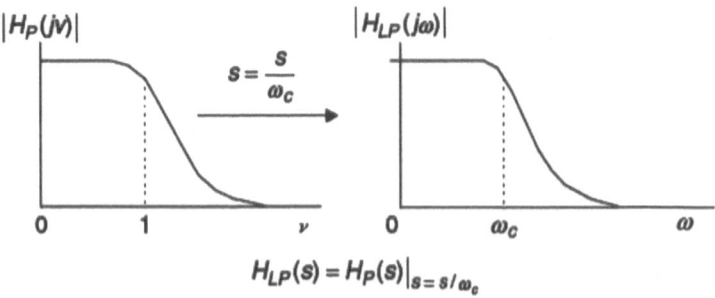

Figure 8.11 Analog lowpass prototype transformation into a lowpass filter.

We compute the gains at $\nu = 0, \nu = 1, \nu = 100, \nu = 10{,}000$ to obtain $1, 1/\sqrt{2}, 0.0995$, and 0.01, respectively. The cutoff frequency gain at $\nu=1$ equals $1/\sqrt{2}$, which is equivalent to $-3\,dB$, and the direct-current (DC) gain is 1. The gain approaches zero when the frequency goes to ν. This verifies that the lowpass prototype is a normalized lowpass filter with a normalized cutoff frequency of 1. Applying the prototype transformation s/ω_c in Figure 8.11, we get an analogue lowpass filter with a cutoff frequency of ω_c as:

$$H(s) = \frac{1}{s/\omega_c + 1} = \frac{\omega_c}{s + \omega_c}. \quad (8.20)$$

We can obtain the analog frequency response by substituting $s = j\omega$ into Equation (8.20), that is,

$$H(j\omega) = \frac{1}{j\omega/\omega_c + 1}.$$

The magnitude response is determined by

$$|H(j\omega)| = \frac{1}{\sqrt{1 + \left(\frac{\omega}{\omega_c}\right)^2}}. \quad (8.21)$$

Similarly, we verify the gains at $\omega = 0$, $\omega = \omega_c$, $\omega = 100\omega_c$, $\omega = 10{,}000\omega_c$ to be 1, $1/\sqrt{2}$, 0.0995, and 0.01 respectively. The filter gain at the cutoff frequency ω_c equal $1/\sqrt{2}$ and the DC gain is 1. The gain approaches zero when $\omega = +\infty$.

We notice that filter gains do not change but that the filter frequency is scaled up by a factor of ω_c. This verifies that the prototype transformation converts the lowpass prototype to the analog lowpass filter with the specified cutoff frequency of ω_c without an effect on the filter gain.

This first-order prototype function is used here for an illustrative purpose.

We will obtain general functions for Butterworth and Chebyshev lowpass prototypes in a later section. The highpass, bandpass, and bandstop filters using the specified lowpass prototype transformation can be easily verified. We review them in Figures 8.12, 8.13, and 8.14, respectively. The transformation from the lowpass prototype to the highpass filter $H_{HP}(s)$ with a cutoff frequency v_c radians/second is given in Figure 8.12, where $x = \omega_c/s$ in the lowpass prototype transformation. The transformation of the lowpass prototype function to a bandpass filter with a center frequency ω_0, a lower cutoff frequency ω_l, and an upper cutoff frequency ω_h in the passband is depicted in Figure 8.13, where $s = (s^2 + \omega_0^2)/=(sW)$ is substituted into the lowpass prototype.

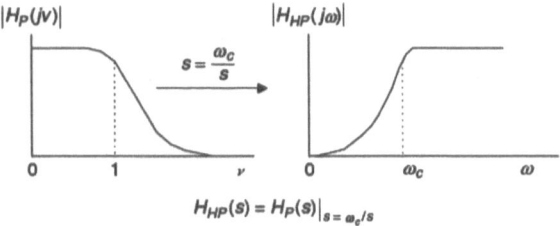

Figure 8.12 Analog lowpass prototype transformation to the highpass filter.

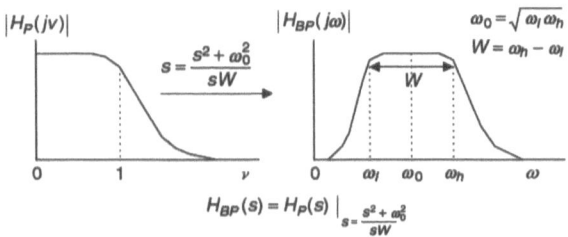

Figure 8.13 Analog lowpass prototype transformation to the bandpass filter.

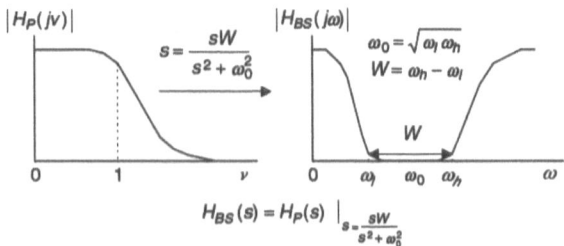

Figure 8.14 Analog lowpass prototype transformation to a bandstop filter

Example 8.6

Given a lowpass prototype

$$H_P(s) = \frac{1}{s+1},$$

Determine each of the following analogue filters and plot their magnitude responses from 0 to 200 radians per second.

1. The highpass filter with a cutoff frequency of 40 radians per second.
2. The bandpass filter with a center frequency of 100 radians per second and bandwidth of 20 radians per second.

Solution:

1. Applying the lowpass prototype transformation by substituting $s = 40/s$ into the lowpass prototype, we have an analog highpass filter as

$$H_{HP}(s) = \frac{1}{\frac{40}{s}+1} = \frac{s}{s+40}.$$

2. Similarly, substituting the lowpass-to-bandpass transformation $s = (s^2 + 100)/(20s)$ into the lowpass prototype leads to:

$$H_{BP}(s) = \frac{1}{\frac{s^2+100}{20s}+1} = \frac{20s}{s^2+20s+100}.$$

After calculation using programming for plotting the magnitude responses for highpass and bandpass filters, figure 8.6 displays the magnitude responses for the highpass filter and bandpass filter, respectively.

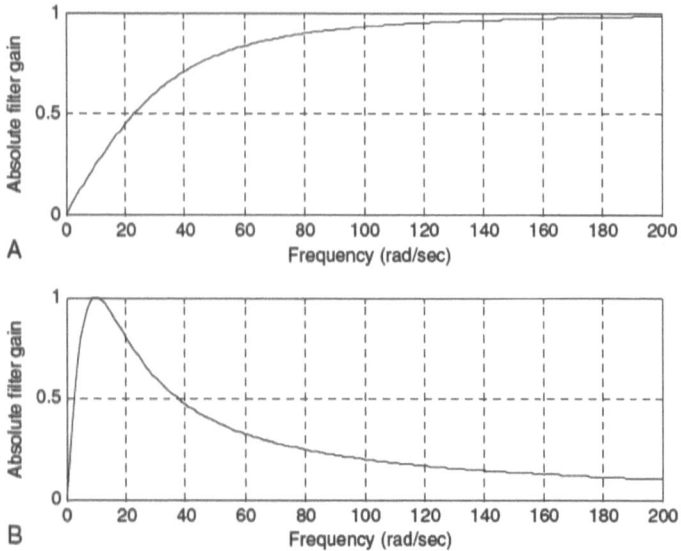

Figure 8.15 Magnitude responses for the analog highpass filter and bandpass filter in

8.3.4 Bilinear Transformation and Frequency Warping

In this subsection, we develop the BLT, which converts an analog filter into a digital filter. We begin by finding the area under a curve using the integration of calculus and the numerical recursive method. The area under the curve is a common problem in early calculus courses. As shown in Figure 8.7, the area under the curve can be determined using the following integration:

$$y(t) = \int_0^t x(t)dt, \tag{8.22}$$

where $y(t)$ (area under the curve) and x(t) (curve function) are the output and input of the analog integrator, respectively, and t is the upper limit of the integration. Applying Laplace transform on Equation (8.22), we have

$$Y(s) = \frac{X(s)}{s}$$

and find the Laplace transfer function as

$$G(s) = \frac{Y(s)}{X(s)} = \frac{1}{s}.$$ (8.7)

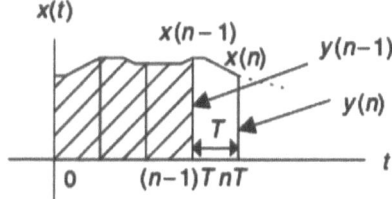

Figure 8.7 Digital integration method to calculate the area under the curve.

Now we examine the numerical integration method shown in Figure 8.7 to approximate the integration of Equation (8.5) using the following difference equation:

$$y(n) = y(n-1) + \frac{x(n) + x(n-1)}{2} T,$$ (8.8)

where T denotes the sampling period. $y(n) = y(nT)$ is the output sample that is the whole area under the curve, while $y(n-1) = y(nT - T)$ is the previous output sample from the integrator indicating the previously computed area under the curve (the shaded area in Figure 8.7). Notice that $x(n) = x(nT)$ and $x(n-1) = x(nT - T)$, sample amplitudes from the curve, are the current input sample and the previous input sample in Equation (8.8). Applying the z-transform on both sides of Equation (8.8) leads to

$$Y(z) = z^{-1}Y(z) + \frac{T}{2}(X(z) + z^{-1}X(z)).$$

Solving for the ratio $Y(z)/X(z)$, we achieve the z-transfer function as

$$H(z) = \frac{Y(z)}{X(z)} = \frac{T}{2}\frac{1+z^{-1}}{1-z^{-1}}. \tag{8:9}$$

Next, comparing Equation (8.9) with Equation (8.7), it follows that

$$\frac{1}{s} = \frac{T}{2}\frac{1+z^{-1}}{1-z^{-1}} = \frac{T}{2}\frac{z+1}{z-1}. \tag{8:10}$$

Solving for s in Equation (8.10) gives the bilinear transformation

$$s = \frac{2}{T}\frac{z-1}{z+1}. \tag{8:11}$$

The BLT method is a mapping or transformation of points from the s-plane to the z- plane. Equation (8.11) can be alternatively written as

$$z = \frac{1+sT/2}{1-sT/2}. \tag{8:12}$$

The general mapping properties are summarized as following:

1. The left-half s-plane is mapped onto the inside of the unit circle of the z-plane.

2. The right-half s-plane is mapped onto the outside of the unit circle of the z-plane.
3. The positive jv axis portion in the s-plane is mapped onto the positive half circle (the dashed-line arrow in Figure 8.8) on the unit circle, while the negative $j\omega$ axis is mapped onto the negative half circle (the dottedline arrow in Figure 8.8) on the unit circle.

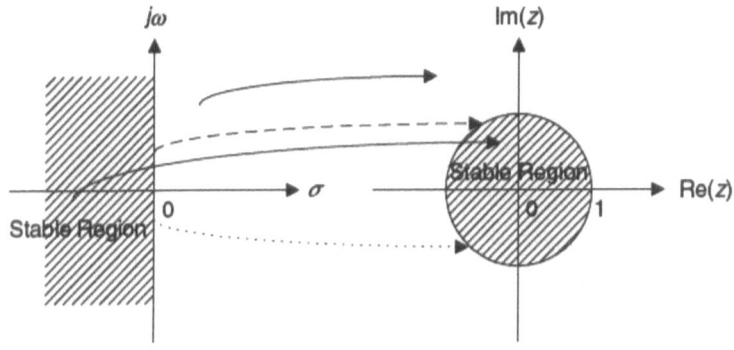

Figure 8.8 Mapping between the s-plane and the z-plane by the bilinear transformation.

Example 8.7

Assuming that T = 2 seconds in Equation (8.12), and given the following points:

1. $s = -1 + j$, on the left half of the s-plane
2. $s = 1 - j$, on the right half of the s-plane
3. $s = j$, on the positive jv on the s-plane
4. $s = -j$, on the negative jv on the s-plane,

Convert each of the points in the s-plane to the z-plane, and verify the mapping properties (1) to (3).

Solution:
Substituting T =2 into Equation (8.12) leads to

$$z = \frac{1+s}{1-s}.$$

We can carry out mapping for each point as follows:

1.

$$z = \frac{1+(-1+j)}{1-(-1+j)} = \frac{j}{2-j} = \frac{1\angle 90°}{\sqrt{5}\angle -26.57°} = 0.4472\angle 116.57°,$$

since $|Z| = 0.4472 < 1$, which is inside the unit circle on the z-plane.

2.

$$z = \frac{1+(1-j)}{1-(1-j)} = \frac{2-j}{j} = \frac{\sqrt{5}\angle -26.57°}{1\angle 90°} = 2.2361\angle -116.57°,$$

since $|Z| = 2.2361 < 1$, which is outside the unit circle on the z-plane.

3.

$$z = \frac{1+j}{1-j} = \frac{\sqrt{2}\angle 45°}{\sqrt{2}\angle -45°} = 1\angle 90°,$$

since $|Z| = 1$ and $\theta = 90°\_$, which is on the positive half circle on the unit circle on the z-plane.

4.

$$z = \frac{1-j}{1-(-j)} = \frac{1-j}{1+j} = \frac{\sqrt{2}\angle -45°}{\sqrt{2}\angle 45°} = 1\angle -90°,$$

Since $|Z| = 1$ and $\theta = -90°$, which is on the negative half circle on the unit circle on the z-plane.

As shown in Example 8.7, the BLT offers conversion of an analog transfer function to a digital transfer function. Example 8.4 shows how to perform the BLT.

Example 8.8.
Given an analog filter whose transfer function is

$$H(s) = \frac{10}{s + 10}$$

Convert it to the digital filter transfer function and difference equation, respectively, when a sampling period is given as T = 0.01 second.

Solution:
Applying the BLT, we have

$$H(z) = H(s)|_{s=\frac{2}{T}\frac{z-1}{z+1}} = \frac{10}{s+10}\Big|_{s=\frac{2}{T}\frac{z-1}{z+1}}$$

Substituting T = 0.01, it follows that

$$H(z) = \frac{10}{\frac{200(z-1)}{z+1} + 10} = \frac{0.05}{\frac{z-1}{z+1} + 0.05} = \frac{0.05(z+1)}{z - 1 + 0.05(z+1)} = \frac{0.05z + 0.05}{1.05z - 0.95}.$$

Finally, we get

$$H(z) = \frac{(0.05z + 0.05)/(1.05z)}{(1.05z - 0.95)/(1.05z)} = \frac{0.0476 + 0.0476z^{-1}}{1 - 0.9048z^{-1}}.$$

Applying the technique in Chapter 6, we yield the difference equation as

$$y(n) = 0.0476x(n) + 0.0476x(n-1) + 0.9048y(n-1)$$

Next, we examine frequency mapping between the s-plane and the z-plane. As illustrated in Figure 8.9, the analog frequency ω_o is marked on the $j\omega$ axis on the s-plane, whereas ω_d is the digital frequency labeled on the unit circle in the z-plane.

We substitute $s = j\omega_a$ and $z = e^{j\omega_d T}$ into the BLT in Equation (8.11) to get

$$j\omega_a = \frac{2}{T} \frac{e^{j\omega_d T} - 1}{e^{j\omega_d T} + 1}.$$

Simplifying Equation (8.13) leads to

$$\omega_a = \frac{2}{T} \tan\left(\frac{\omega_d T}{2}\right).$$

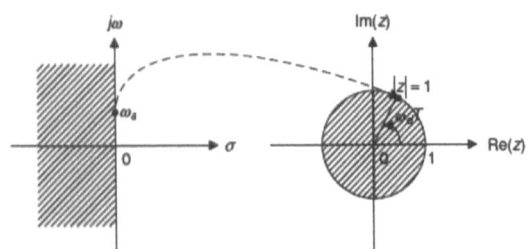

Figure 8.9 Frequency mapping from the analog domain to the digital domain.

Equation (8.14) explores the relation between the analog frequency on the $j\omega$ axis and the corresponding digital frequency ω_d on the unit circle. We can also write its inverse as:

$$\omega_d = \frac{2}{T}\tan^{-1}\left(\frac{\omega_a T}{2}\right).$$
(8:15)

The range of the digital frequency ω_d is from 0 radian per second to the folding frequency $\omega_s/2$ per second, where ω_s is the sampling frequency in radians per second. We make a plot of Equation (8.14) in Figure 8.10.

From Figure 8.10 when the digital frequency range $0 \le \omega_d \le 0.25\omega_s$ is mapped to the analog frequency range $0 \le \omega_a \le 0.32\omega_s$, the transformation appears to be linear; however, when the digital frequency range $02.5\omega_s \le \omega_d \le 0.5\omega_s$ is mapped to the analog frequency range for $\omega_a > 0.32\omega_s$, the transformation is nonlinear. The analog frequency range for $\omega_a > 0.32\omega_s$ is compressed into the digital frequency range $0.25\omega_s \le \omega_d \le 0.5\omega_s$. This nonlinear frequencymapping effect is called frequency warping. We must incorporate the frequency warping into the IIR filter design. The following example will illustrate the frequency warping effect in the BLT.

8.3.5 Bilinear Transformation Design Procedure

Now we can summarize the BLT design procedure.

1. Given the digital filter frequency specifications, prewarp the digital frequency specifications to the analog frequency specifications.

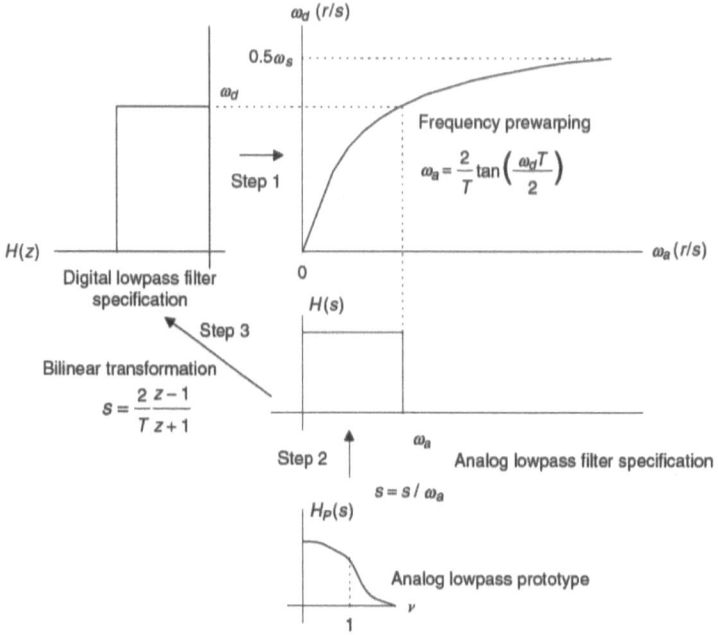

Figure 8.11 representation of IIR filter design using the bilinear transformation.

For the lowpass filter and highpass filter:

$$\omega_a = \frac{2}{T} \tan\left(\frac{\omega_d T}{2}\right). \tag{8.18}$$

For the bandpass filter and bandstop filter:

$$\omega_{al} = \frac{2}{T} \tan\left(\frac{\omega_l T}{2}\right), \quad \omega_{ah} = \frac{2}{T} \tan\left(\frac{\omega_h T}{2}\right), \tag{8.19}$$

where

$$\omega_0 = \sqrt{\omega_{al}\omega_{ah}}, \quad W = \omega_{ah} - \omega_{al}$$

2. Perform the prototype transformation using the lowpass prototype $Hp(s)$.

From lowpass to lowpass: $H(s) = H_P(s)|_{s=\frac{s}{\omega_a}}$

From lowpass to highpass: $H(s) = H_P(s)|_{s=\frac{\omega_a}{s}}$

From lowpass to bandpass: $H(s) = H_P(s)|_{s=\frac{s^2+\omega_0^2}{sW}}$

From lowpass to bandstop: $H(s) = H_P(s)|_{s=\frac{sW}{s^2+\omega_0^2}}$

3. Substitute the BLT to obtain the digital filter

$$H(z) = H(s)|_{s=\frac{2}{T}\frac{z-1}{z+1}}.$$

Example 8.9

Design a digital lowpass Butterworth filter with the following specifications:

1. 3 dB attenuation at the passband frequency of 1.5 kHz
2. 10 dB stopband attenuation at the frequency of 3 kHz
3. Sampling frequency of 8,000 Hz.

Solution:
First, we obtain the digital frequencies in radians per second:

$$\omega_{dp} = 2\pi f = 2\pi(1500) = 3000\pi \, rad = sec$$
$$\omega_{ds} = 2\pi f = 2\pi(3000) = 6000\pi \, rad = sec$$
$$T = 1/f_s = 1/8000 \, sec$$

Following the steps of the design procedure,

1. We apply the warping equation as

$$\omega_{ap} = \frac{2}{T}\tan\left(\frac{\omega_d T}{2}\right) = 16000 \times \tan\left(\frac{3000\pi/8000}{2}\right) = 1.0691 \times 10^4 \text{ rad/sec.}$$

$$\omega_{as} = \frac{2}{T}\tan\left(\frac{\omega_d T}{2}\right) = 16000 \times \tan\left(\frac{6000\pi/8000}{2}\right) = 3.8627 \times 10^4 \text{ rad/sec.}$$

We then find the lowpass prototype specifications using Table 8.6 as follows:

$$v_s = \omega_{as}/\omega_{ap} = 3.8627 \times 10^4/(1.0691 \times 10^4)$$
$$= 3.6130 \text{ rad/sec and } A_s = 10 \text{ dB.}$$

The filter order is computed as

$$\varepsilon^2 = 10^{0.1 \times 3} - 1 = 1$$
$$n = \frac{\log_{10}(10^{0.1 \times 10} - 1)}{2 \cdot \log_{10}(3.6130)} = 0.8553.$$

2. Rounding n up, we choose n ¼ 1 for the lowpass prototype. From Table 8.3, we have

$$H_P(s) = \frac{1}{s+1}.$$

Applying the prototype transformation (lowpass to lowpass) yields the analog filter

$$H(s) = H_P(s)|_{\frac{s}{\omega_{ap}}} = \frac{1}{\frac{s}{\omega_{ap}}+1} = \frac{\omega_{ap}}{s+\omega_{ap}} = \frac{1.0691 \times 10^4}{s+1.0691 \times 10^4}.$$

3. Finally, using the BLT, we have

$$H(z) = \frac{1.0691 \times 10^4}{s + 1.0691 \times 10^4}\bigg|_{s=16000(z-1)/(z+1)}.$$

Substituting the BLT leads to

$$H(z) = \frac{1.0691 \times 10^4}{\left(16000\frac{z-1}{z+1}\right) + 1.0691 \times 10^4}.$$

To simplify the algebra, we divide both numerator and denominator by 16000 to get

$$H(z) = \frac{0.6682}{\left(\frac{z-1}{z+1}\right) + 0.6682}.$$

Then multiplying $(z + 1)$ to both numerator and denominator leads to

$$H(z) = \frac{0.6682(z+1)}{(z-1) + 0.6682(z+1)} = \frac{0.6682z + 0.6682}{1.6682z - 0.3318}.$$

Dividing both numerator and denominator by (1:6682 _z) leads to

$$H(z) = \frac{0.4006 + 0.4006z^{-1}}{1 - 0.1989z^{-1}}.$$

Figure 8.16 describes the filter frequency responses.

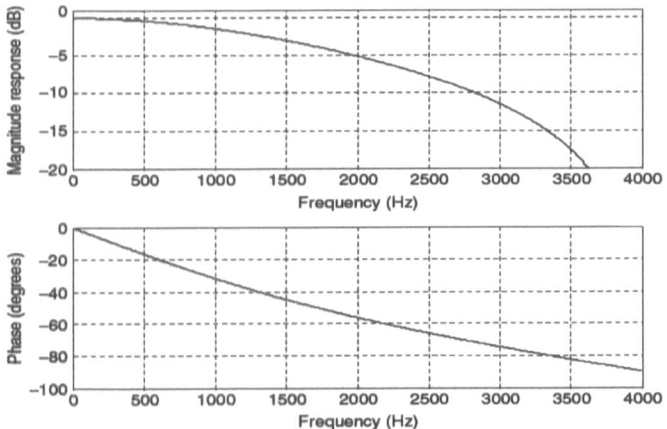

Figure 8.16 Frequency responses of the designed digital filter for Example 8.7.

8.4.6 Impulse Invariant Design Method

We illustrate the concept of the impulse invariant design in Figure 8.27. Given the transfer function of a designed analog filter, an analog impulse response can be easily found by the inverse Laplace transform of the transfer function.

To replace the analog filter by the equivalent digital filter, we apply an approximation in time domain in which the digital impulse response must be equivalent to the analog impulse response. Therefore, we can sample the analog impulse response to get the digital impulse response and take the z-transform of the sampled analog impulse response to obtain the transfer function of the digital filter. The analog impulse response can be achieved by taking the inverse Laplace transform of the analog filter H(s), that is,

$$h(t) = L^{-1}(H(s)).$$

Now, if we sample the analog impulse response with a sampling interval of T and use T as a scale factor, it follows that

$$T \cdot h(n) = T \cdot h(t)|_{t=nT}, \ n \geq 0.$$

Taking the z-transform on both sides of Equation (8.38) yields the digital filter as

$$H(z) = Z[T \cdot h(n)].$$

The effect of the scale factor T in Equation (8.38) can be explained as follows. We approximate the area under the curve specified by the analog impulse function h(t) using a digital sum given by

$$\text{area} = \int_0^\infty h(t)dt \approx T \cdot h(0) + T \cdot h(1) + T \cdot h(2) + \cdots$$

Note that the area under the curve indicates the DC gain of the analog filter while the digital sum in Equation (8.40) is the DC gain of the digital filter.

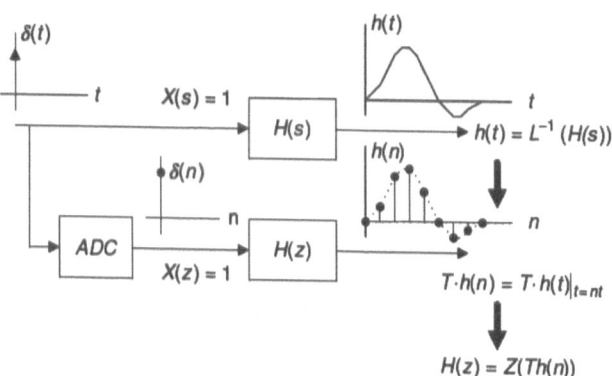

Figure 8.27 Impulse invariant design method.

The rectangular approximation is used, since each sample amplitude is multiplied by the sampling interval T. Due to the interval size for approximation in practice, we cannot guarantee that the digital sum has exactly the same value as the one from the integration unless the sampling interval T in Equation (8.40) approaches zero. This means that the higher the sampling rate— that is, the smaller the sampling interval—the more accurately the digital filter gain matches the analog filter gain. Hence, in practice, we need to further apply gain scaling for adjustment if it is a requirement. We look at the following examples.

Example 8.9
Consider the following Laplace transfer function:

$$H(s) = \frac{2}{s+2}$$

Determine $H(z)$ using the impulse invariant method if the sampling rate $fs = 10$ Hz.

Solution
Taking the inverse Laplace transform of the analog transfer function, the impulse response is found to be

$$h(t) = L^{-1}\left[\frac{2}{s+2}\right] = 2e^{-2t}u(t).$$

Sampling the impulse response $h(t)$ with $T = 1/fs = 0.1$ second, we Have

$$Th(n) = T2e^{-2nT}u(n) = 0.2e^{-0.2n}u(n).$$

Using the z-transform table in Chapter 5, we yield

$$Z[e^{-an}u(n)] = \frac{z}{z - e^{-a}}.$$

And noting that

$$H(z) = \frac{0.2z}{z - 0.8187} = \frac{0.2}{1 - 0.8187z^{-1}}.$$

Problems

8.1. Design a 3-tap FIR lowpass filter with a cutoff frequency of 1,500 Hz and a sampling rate of 8,000 Hz using :

  i. rectangular window function
  ii. Hamming window function.
  Determine the transfer function and difference equation of the designed FIR system, and compute and plot the magnitude frequency response for $V = 0$, $p=4$, $p=2$, $3p=4$; and p radians.

8.2. Design a 3-tap FIR highpass filter with a cutoff frequency of 1,600 Hz and a sampling rate of 8,000 Hz using:

  i. rectangular window function
  ii. Hamming window function.
  Determine the transfer function and difference equation of the designed FIR system, and compute and plot the magnitude frequency response for $V ¼ 0$, $p=4$, $p=2$, $3p=4$; and p radians.

8.3. Design a 5-tap FIR bandpass filter with a lower cutoff frequency of 1,600 Hz, an upper cutoff frequency of 1,800 Hz, and a sampling rate of 8,000 Hz using :

  i. rectangular window function
  ii. Hamming window function.
  Determine the transfer function and difference equation of the designed FIR system, and compute and plot the magnitude frequency response for $V ¼ 0$, $p=4$, $p=2$, $3p=4$; and p radians.

8.4. Design a 5-tap FIR band reject filter with a lower cutoff frequency of 1,600 Hz, an upper cutoff frequency of 1,800 Hz, and a sampling rate of 8,000 Hz using

  i. rectangular window function
  ii. Hamming window function.
  Determine the transfer function and difference equation of the designed FIR system, and compute and plot the magnitude frequency response for V ¼ 0, p=4, p=2,

8.5. Given an FIR lowpass filter design with the following specifications:

Passband = 0–800 Hz,
Stopband = 1,200–4,000 Hz
Passband ripple = 0.1 dB
Stopband attenuation = 40 dB
Sampling rate = 8,000 Hz,
determine the following:
  i. window method
  ii. length of the FIR filter
  iii. cutoff frequency for the design equation.

8.6. Given an FIR highpass filter design with the following specifications:
Passband = 0–1,500 Hz Stopband = 2,000–4,000 Hz
Passband ripple = 0.02 dB Stopband attenuation = 60 dB Sampling rate = 8,000 Hz,
determine the following:

  i. window method
  ii. length of the FIR filter
  iii. cutoff frequency for the design equation.

8.7. Given an FIR bandpass filter design with the following specifications:
Lower cutoff frequency = 1,500 Hz Lower transition width = 600 Hz Upper cutoff frequency = 2,300 Hz Upper transition width = 600 Hz Passband ripple = 0.1 dB Stopband attenuation = 50 dB Sampling rate = 8,000 Hz,
determine the following:
window method
length of the FIR filter cutoff frequencies for the design equation.

8.8. Given an FIR bandstop filter design with the following specifications:
Lower passband = 0–1,200 Hz Stopband = 1,600–2,000 Hz Upper passband = 2,400–4,000 Hz Passband ripple = 0.05 dB Stopband attenuation = 60 dB Sampling rate = 8,000 Hz,
determine the following:
window method
length of the FIR filter cutoff frequencies for the design equation.

8.9. Given an FIR system
$$H(z) = 0.25 - 0.5z^{-1} + 0.25z^{-2},$$
realize $H(z)$ using each of the following specified methods:
i. transversal form, and write the difference equation for implementation
ii. linear phase form, and write the difference equation for implementation

8.10. Given an FIR filter transfer function
$$H(z) = 0.2 + 0.5z^{-1} - 0.3z^{-2} + 0.5z^{-3} + 0.2z^{-4},$$

8.4. Design a 5-tap FIR band reject filter with a lower cutoff frequency of 1,600 Hz, an upper cutoff frequency of 1,800 Hz, and a sampling rate of 8,000 Hz using

  i. rectangular window function
  ii. Hamming window function.
  Determine the transfer function and difference equation of the designed FIR system, and compute and plot the magnitude frequency response for V ¼ 0, p=4, p=2,

8.5. Given an FIR lowpass filter design with the following specifications:

  Passband = 0–800 Hz,
  Stopband = 1,200–4,000 Hz
  Passband ripple = 0.1 dB
  Stopband attenuation = 40 dB
  Sampling rate = 8,000 Hz,
  determine the following:
  i. window method
  ii. length of the FIR filter
  iii. cutoff frequency for the design equation.

8.6. Given an FIR highpass filter design with the following specifications:
  Passband = 0–1,500 Hz Stopband = 2,000–4,000 Hz
  Passband ripple = 0.02 dB Stopband attenuation = 60 dB Sampling rate = 8,000 Hz,
  determine the following:

  i. window method
  ii. length of the FIR filter
  iii. cutoff frequency for the design equation.

8.7. Given an FIR bandpass filter design with the following specifications:
Lower cutoff frequency = 1,500 Hz Lower transition width = 600 Hz Upper cutoff frequency = 2,300 Hz Upper transition width = 600 Hz Passband ripple = 0.1 dB Stopband attenuation = 50 dB Sampling rate = 8,000 Hz,
determine the following:
window method
length of the FIR filter cutoff frequencies for the design equation.

8.8. Given an FIR bandstop filter design with the following specifications:
Lower passband = 0–1,200 Hz Stopband = 1,600–2,000 Hz Upper passband = 2,400–4,000 Hz Passband ripple = 0.05 dB Stopband attenuation = 60 dB Sampling rate = 8,000 Hz,
determine the following:
window method
length of the FIR filter cutoff frequencies for the design equation.

8.9. Given an FIR system
$$H(z) = 0.25 - 0.5z^{-1} + 0.25z^{-2},$$
realize $H(z)$ using each of the following specified methods:
i. transversal form, and write the difference equation for implementation
ii. linear phase form, and write the difference equation for implementation.

8.10. Given an FIR filter transfer function
$$H(z) = 0.2 + 0.5z^{-1} - 0.3z^{-2} + 0.5z^{-3} + 0.2z^{-4},$$

perform the linear phase FIR filter realization, and write the difference equation for implementation.

Determine the transfer function for a 5-tap FIR lowpass filter with a lower cutoff frequency of 2,000 Hz and a sampling rate of 8,000 Hz using the frequency sampling method.

Determine the transfer function for a 5-tap FIR highpass filter with a lower cutoff frequency of 3,000 Hz and a sampling rate of 8,000 Hz using the frequency sampling method.

Given the following specifications:

a 7-tap FIR bandpass filter

a lower cutoff frequency of 1,500 Hz and an upper cutoff frequency of 3,000 Hz

    a sampling rate of 8,000 Hz the frequency sampling design method, determine the transfer function.

8.11. Given the following specifications:

a 7-tap FIR band reject filter a lower cutoff frequency of 1,500 Hz and an upper cutoff frequency of 3,000 Hz a sampling rate of 8,000 Hz, the frequency sampling design method, determine the transfer function.

8.12. In a speech recording system with a sampling rate of 10,000 Hz, the speech is corrupted by broadband random noise. To remove the random noise while preserving speech information, the following specifications are given:

Speech frequency range 0–3,000 kHz Stopband range 4,000–5,000 Hz Passband ripple 0.1 dB

Stopband attenuation 45 dB FIR filter with Hamming window.

Determine the FIR filter length (number of taps) and the cutoff frequency; use MATLAB to design the filter; and plot the frequency response.

8.13. Given a speech equalizer shown in Figure 8.13 to compensate midrange frequency loss of hearing:
   Sampling rate 8,000 Hz
   Bandpass FIR filter with Hamming window
   Frequency range to be emphasized = 500–2,000 Hz
   Lower stopband = 0–1,000 Hz
   Upper stopband = 2,500–4,000 Hz
   Passband ripple = 0.1 dB
Stopband attenuation = 45 dB,

determine the filter length and the lower and upper cutoff frequencies.

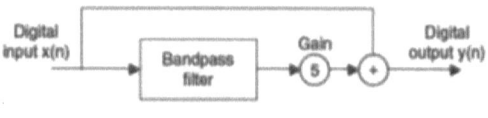

Figure 8.13

8.14. Given an analog filter with the transfer function

$$H(s) = \frac{1000}{s + 1000}$$

convert it to the digital filter transfer function and difference equation using the BLT if the DSP system has a sampling period of T = 0.001 second.

8.15 The lowpass filter with a cutoff frequency of 1 rad/sec is given as

$$H_p(s) = \frac{1}{s+1}$$

Use $H_p(s)$ and the BLT to obtain a corresponding IIR digital lowpass filter with a cutoff frequency of 30 Hz, assuming a sampling rate of 200 Hz.

8.16. The normalized lowpass filter with a cutoff frequency of 1 rad/sec is given as

$$H_p(s) = \frac{1}{s+1}$$

Use $H_p(s)$ and the BLT to obtain a corresponding IIR digital highpass filter with a cutoff frequency of 30 Hz, assuming a sampling rate of 200 Hz.

8.17. Consider the normalized lowpass filter with a cutoff frequency of 1 rad/sec:

$$H_p(s) = \frac{1}{s+1}$$

Use Hp(s) and the BLT to design a corresponding IIR digital notch (bandstop) filter with a lower cutoff frequency of 20 Hz, an upper cutoff frequency of 40 Hz, and a sampling rate of 120 Hz.

8.18 Consider the following normalized lowpass filter with a cutoff frequency of 1 rad/sec:

$$H_p(s) = \frac{1}{s+1}$$

Use Hp(s) and the BLT to design a corresponding IIR digital bandpass filter with a lower cutoff frequency of 15 Hz, an upper cutoff frequency of 25 Hz, and a sampling rate of 120 Hz.

8.19. Design a first-order digital lowpass Butterworth filter with a cutoff requency of 1.5 kHz and a passband ripple of 3 dB at a sampling frequency of 8,000 Hz.

Determine the transfer function and difference equation.

8.20 Design a second-order digital lowpass Butterworth filter with a cutoff frequency of 1.5 kHz and a passband ripple of 3 dB at a sampling frequency of 8,000 Hz.

Determine the transfer function and difference equation.

8.21 Consider the following Laplace transfer function: Determine H(z) and the difference equation using the impulse invariant method if the sampling rate fs =10 Hz.

$$H_p(s) = \frac{1}{s^2 + 3s + 2}$$

8.22. A speech sampled at 8,000 Hz is corrupted by a sine wave of 360 Hz. Design a notch filter to remove the noise with the following specifications:
Chebyshev notch filter
Center frequency: 360 Hz
Bandwidth: 60 Hz
Passband and ripple: 0.5 dB
Stopband attenuation: 5 dB at 355 Hz and 365 Hz, respectively.

Determine the transfer function and difference equation.

Chapter Nine

Wavelet Transform

Learning Outcomes of this Chapter

After successful completion of this chapter students will be able to:

1. understand the windowed Fourier transform and difference between windowed Fourier transform and wavelet transform.
2. understand wavelet basis and characterize continuous and discrete wavelet transforms.
3. Understand multi resolution analysis and identify various wavelets and evaluate their frequency resolution properties.
4. implement discrete wavelet transforms for few application such as denoising for IoT networks, signal detection, ..etc.

9.1 Introduction

The wavelet transform is a mathematical tool developed mainly since the middle of the 1980's. It is efficient for local analysis of nonstationary and fast transient wide-band signals. The wavelet transform is a mapping of a time signal to the time-scale joint representation that is *like* the short-time Fourier transform, the Wigner distribution and the ambiguity function. The temporal aspect of the signals is preserved. The wavelet transform provides multiresolution analysis with dilated windows. The high frequency analysis is done using narrow windows and the low frequency analysis is done using wide windows.

The base of the wavelet transforms, the wavelets, are generated from a basic wavelet function by dilations and translations. They satisfy an admissible condition so that the original signal can be reconstructed by the inverse wavelet transform. The wavelets also satisfy the regularity condition so that the wavelet coefficients decrease fast with decreases of the scale. The wavelet transform is not only local in time but also in frequency.

Wavelet Transform (WT) is particularly suitable for application of non-stationary signals which may instantaneously vary in time. Primarily, the received signal is divided into different frequency components using wavelets. The basis function of WT is scaled based on frequency and a subset of small waves (known as mother wavelet) is used for implementing WT. The mother wavelet is a time-varying window function used for decomposition of $x(i)$ into weighted sets of scaled versions of $y(i)$. Consequently, using wavelet transform in signal processing is the process of the partial transformation of the spatial domain and the frequency domain, in order to get useful information accurately from it though corrupted with noise.

Since different frequency levels are used for WT, it is quite convenient for analyzing the signal characteristics at different frequencies and detecting removing corrupting noise. Broadly, there are two types of WT, Continuous Wavelet Transform (CWT) and Discrete Wavelet Transform (DWT).

9.2 Continuous Wavelet Transform

Continuous *wavelets* transform (CWT) measures the congruence between an analyzing function and actual signal by calculating the inner product and then integrating the product. The mother wavelet window function can be shifted and moved over the time-axis by changing scale and position parameters, thereby including different frequency components at the different locations.

The wavelet *transforms* $W_x(b, a)$ of a continuous-time signal $x(t)$ is defined as:

$$W_x(b, a) = \frac{1}{|a|^{\frac{1}{2}}} \int_{-\infty}^{\infty} x(t) \psi^* \left(\frac{t-b}{a}\right) dt \quad (9.1)$$

Thus, the wavelet transform is computed as the inner product of $x(t)$ and translated and scaled versions of a single function $\psi(t)$, the so-called wavelet. If we consider $\psi(t)$ to be a bandpass impulse response, then the wavelet analysis can be understood as a bandpass analysis. By varying the scaling parameter a the center frequency and the bandwidth of the bandpass are influenced. The variation of b simply means a translation in time, so that for a fixed a the transform (9.1) can be seen as a convolution of $x(t)$ with the time-reversed and scaled wavelet:

$$W_x(t, a) = |a|^{-\frac{1}{2}} x(t) * \psi_a(t), \qquad \psi_a(t) = \psi(t)^* (\tfrac{-t}{a}).$$

The perfector $|a|^{-\frac{1}{2}}$ is introduced in order to ensure that all scaled functions $|a|^{-\frac{1}{2}} \psi(t)^*(\tfrac{-t}{a})$ with $a \in R$ have the same energy.

Since the analysis function $\psi(t)$ is scaled and not modulated like the kernel of the STFT, a wavelet analysis is often called a time-scale analysis rather than a time-frequency analysis. However, both are naturally related to each other by the bandpass interpretation. Figure 9.1 shows examples of the kernels of the STFT and the wavelet transform. As we can see, a variation of the time delay b and/or of the scaling parameter a has no effect on the form of the transform kernel of the wavelet transform. However, the time and frequency resolution of the wavelet transform depends on a. For high analysis frequencies (small a) we have good time localization but poor frequency resolution. On the other hand, for low analysis frequencies, we have good frequency but poor time resolution. While the STFT is a constant bandwidth analysis, the wavelet analysis can be understood as a constant-Q or octave analysis.

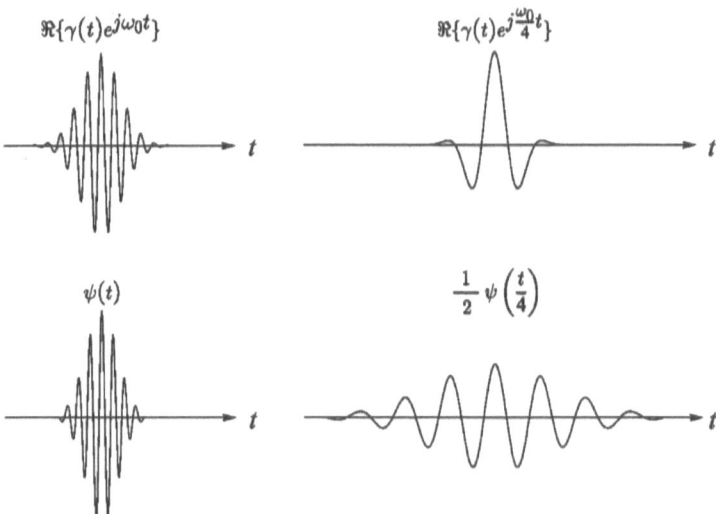

Figure 9.1. Comparison of the analysis kernels of the short-time Fourier transform (top, the real part is shown) and the wavelet transform (bottom, real wavelet) for high and low analysis frequencies.

When using a transform in order to get better insight into the properties of a signal, it should be ensured that the signal can be perfectly reconstructed from its representation. Otherwise, the representation may be completely or partly meaningless. For the wavelet transform the condition that must be met in order to ensure perfect reconstruction is

$$C_\psi = \int_{-\infty}^{\infty} \frac{|\Psi(\omega)|^2}{|\omega|} d\omega < \infty \qquad (9.2)$$

where $\Psi(\omega)$ denotes the Fourier transform of the wavelet. This condition is known as the admissibility condition for the wavelet $\psi(t)$. The proof of (9.2) will be given in Section 9.3.

Obviously, in order to satisfy (9.2) the wavelet must satisfy

$$\Psi(0) = \int_{-\infty}^{\infty} \psi(t)\, dt = 0 \tag{9.3}$$

Moreover, $|\Psi(\omega)|$ must decrease rapidly for $[\omega] \to 0$ and for $|\omega| \to \infty$. That is, $\psi(t)$. must be a bandpass impulse response. Since a bandpass impulse response looks like a small wave, the transform is named wavelet transform.

Calculation of the Wavelet Transform from the Spectrum X(w). Using the abbreviation

$$\psi_{b,a}(t) = |a|^{-\frac{1}{2}} \psi\left(\frac{t-b}{a}\right) \tag{9.4}$$

the integral wavelet transform introduced by equation (9.1) can also be written as

$$W_x(b, a) = \langle x, \psi_{b,a} \rangle \tag{9.5}$$

With the correspondences $X(w) \leftrightarrow x(t)$ and $\Psi(\omega) \leftrightarrow \psi(t)$, and the time and frequency shift properties of the Fourier transform, we obtain:

$$\Psi_{b,a}(\omega) = |a|^{-\frac{1}{2}} e^{-j\omega b}\, \Psi(a\omega)$$

$$\psi_{b,a}(t) = |a|^{-\frac{1}{2}} \psi\left(\frac{t-b}{a}\right) \tag{9.6}$$

By making use of Parseval's relation we finally get

$$W_x(b,a) = \frac{1}{2\pi} \langle X, \Psi_{b,a} \rangle$$

$$= |a|^{-\frac{1}{2}} \int_{-\infty}^{\infty} X(\omega) \Psi^*(a\omega) e^{-j\omega b} dw \tag{9.7}$$

Equation (9.7) states that the wavelet transform can also be calculated by means of an inverse Fourier transform from the windowed spectrum $X(\omega)\Psi^*(a\omega)$.

To recover the original signal $x(t)$, the first inverse continuous wavelet transform can be exploited:

$$x(t) = C_\psi^{-1} \int_{-\infty}^{\infty} \int_{-\infty}^{\infty} W_x(a,b) \frac{1}{|a|^{\frac{1}{2}}} \bar{\psi}\left(\frac{t-b}{a}\right) db \frac{da}{a^2}$$

$\bar{\psi}(t)$ is the dual function of $\psi(t)$ and

$$C_\psi = \int_{-\infty}^{\infty} \frac{\bar{\hat{\psi}}(\omega)\hat{\psi}(\omega)}{|\omega|}$$

is admissible constant, where hat means Fourier transform operator. Sometimes, $\hat{\psi}(\omega) = \psi(t)$, then the admissible constant becomes:

$$C_\psi = \int_{-\infty}^{\infty} \frac{|\bar{\hat{\psi}}(\omega)|^2}{|\omega|} d\omega$$

Traditionally, this constant is called wavelet admissible constant. A wavelet whose admissible constant satisfies

$$0 < C_\psi < \infty$$

is called an admissible wavelet. An admissible wavelet implies that $\hat{\psi}(\omega) = 0$ so that an admissible wavelet must integrate to zero.

9.3 Time-Frequency Resolution

To describe the time-frequency resolution of the wavelet transform we consider the time-frequency window associated with the wavelet. According to the time-frequency resolution where the shift and modulation principle of the Fourier transform is applied.

The correspondence center (t_0, ω_0) and the radii Δt and Δw of the window are calculated and gives.

$$t_0 = \int_{-\infty}^{\infty} t \cdot \frac{|\psi(t)|^2}{\|\psi\|^2} \, dt, \tag{9.8}$$

$$\omega_0 = \int_{-\infty}^{\infty} \omega \cdot \frac{|\Psi(\omega)|^2}{\|\Psi\|^2} \, d\omega \tag{9.9}$$

and

$$\Delta_t = \left(\int_{-\infty}^{\infty} (t - t_0)^2 \cdot \frac{|\psi(t)|^2}{\|\psi\|^2} \, dt \right)^{\frac{1}{2}}, \tag{9.10}$$

$$\Delta_\omega = \left(\int_{-\infty}^{\infty} (\omega - \omega_0)^2 \cdot \frac{|\Psi(\omega)|^2}{\|\Psi\|^2} \, d\omega \right)^{\frac{1}{2}}. \tag{9.11}$$

For the center and the radii of the scaled function $\psi\left(\frac{t}{a}\right) \leftrightarrow |a|\hat{\psi}(a\omega)$ we have $\{a. t_0, + \frac{1}{a}\omega_0\}$ and $\{a. \Delta_t, + \frac{1}{a}\Delta_\omega\}$, respectively. This means that the wavelet transform $W_x(b,a)$ provides information on a signal $x(t)$ its spectrum $X(w)$ in the time-frequency window

$$[b+a\cdot t_0 - a\cdot \Delta_t,\ b+a\cdot t_0 + a\cdot \Delta_t] \times [\frac{\omega_0}{a} - \frac{\Delta_\omega}{a},\ \frac{\omega_0}{a} + \frac{\Delta_\omega}{a}]. \quad (9.12)$$

The area $4\Delta_t\Delta_\omega$, is independent of the parameters a and b; it is determined only by the used wavelet $\psi(t)$. The time window narrows when a becomes small, and it widens when a becomes large. On the other hand, the frequency window becomes wide when a becomes small, and it becomes narrow when a becomes large. As mentioned earlier, a short analysis window leads to good time resolution on the one hand, but on the other to poor frequency resolution. Accordingly, a long analysis window yields good frequency resolution but poor time resolution. Figure 9.2 illustrates the different resolutions of the short-time Fourier transform and the wavelet transform.

Affine Invariance. Equation (9.1) shows that if the signal is scaled ($x(t) \rightarrow x(t/c)$), the wavelet representation $W_x(b,a)$ is scaled as well; except this, $W_x(b,a)$ undergoes no other modification. For this reason, we also speak of an *afine invariant transform*. Furthermore, the wave let transform is *translation invariant*, i.e. a shift of the signal ($x(t) \rightarrow x(t-t_0)$) leads to a shift of the wavelet representation $W_x(b,a)$ by t_0, but $W_x(b,a)$ undergoes no other modification.

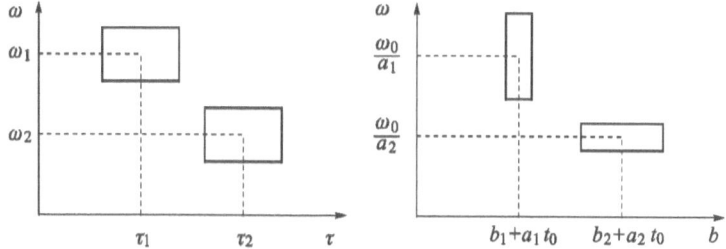

Figure 9.2. Resolution of the short-time Fourier transform (left) and the wavelet transform (right).

9.4 Wavelet Series

9.4.1 Dyadic Sampling

In this section, we consider the reconstruction from discrete values of the wavelet transform. The following dyadically arranged sampling points are used:

$$a_m = 2^m, \qquad b_{mn} = a_m\, n\, T = 2^m nT, \qquad (9.13)$$

This yields the values $W_x(b_{mn}, a_m) = W_x(2^m nT, 2^m)$. Figure 8.6 shows the sampling grid.

Using the abbreviation

$$\begin{aligned}\psi_{mn}(t) &= |a_m|^{-\frac{1}{2}} \cdot \psi\left(\frac{t - b_{mn}}{a_m}\right) \\ &= 2^{-\frac{m}{2}} \cdot \psi(2^{-m}t - nT),\end{aligned} \qquad (9.14)$$

we may write the wavelet analysis as

$$\mathcal{W}_x(b_{mn}, a_m) = \mathcal{W}_x(2^m nT, 2^m) = \langle x, \psi_{mn} \rangle. \quad (9.15)$$

The values $\{W_x(2^m nT, 2^m), m, n \in Z\}$ form the representation of $x(t)$ with respect to the wavelet $\psi(t)$ and the chosen grid.

Of course, we cannot assume that any set $\psi_{mn}(t), m, n \in Z$ allows reconstruction of all signals $x(t) \in L_2(R)$. For this a dual set $\hat{\psi}_{mn}(t), m, n \in Z$ must exist, and both sets must span $L_2(R)$. The dual setn eed not necessarily be built from wavelets. However, we are only interested in the case where $\hat{\psi}_{mn}(t)$ is derived as

$$\tilde{\psi}_{mn}(t) = 2^{-\frac{m}{2}} \cdot \tilde{\psi}(2^{-m} t - nT), \quad m, n \in \mathbb{Z} \quad (9.16)$$

from a dual wavelet $\psi(t)$. If both sets $\psi_{mn}(t)$ and $\hat{\psi}_{mn}(t)$ with $m, n \in Z$ span the space $L_2(R)$., any $x(t) \in L_2(R)$. may be written as

$$x(t) = \sum_{m=-\infty}^{\infty} \sum_{n=-\infty}^{\infty} \langle x, \psi_{mn} \rangle \tilde{\psi}_{mn}(t). \quad (9.17)$$

Alternatively, we may write $x(t)$ a

$$x(t) = \sum_{m=-\infty}^{\infty} \sum_{n=-\infty}^{\infty} \left\langle x, \tilde{\psi}_{mn} \right\rangle \psi_{mn}(t). \quad (9.18)$$

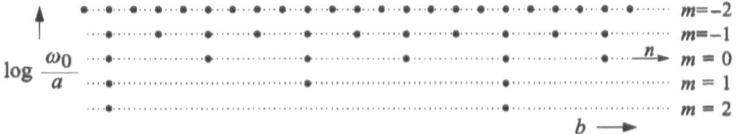

Figure 9.3. Dyadic sampling of the wavelet transform.

For a given wavelet $x(t)$, the possibility of perfect reconstruction is dependent on the sampling interval T. If T is chosen very small (oversampling), the values $W_x(2^m nT, 2^m), m, n \in Z$ are highly redundant, and reconstruction is very easy. Then the functions $\psi_{mn}(t), n \in Z$ are linearly dependent, and an infinite number of dual sets $\hat{\psi}_{mn}(t)$ exists. The question of whether a dual set $\hat{\psi}_{mn}(t)$ exists at all can be answered by checking two frame bounds' A and B. It can be shown that the existence of a dual set and the completeness are guaranteed if the stability condition

$$A \|x\|^2 \leq \sum_{m=-\infty}^{\infty} \sum_{n=-\infty}^{\infty} |\langle x, \psi_{mn} \rangle|^2 \leq B \|x\|^2 \tag{9.19}$$

with the frame bounds $0 < A \leq B < \infty$ is satisfied. In the case of a tight frame, $A = B$, perfect reconstruction with $\psi_{mn}(t) = \hat{\psi}_{mn}(t)$ possible. This is also true if the samples $W_x(2^m nT, 2^m)$ contain redundancy, that is, if the functions $\psi_{mn}(t), m, n \in Z$ are linearly dependent. The tighter the frame bounds are, the smaller is the reconstruction error if the reconstruction is carried out according to

$$\hat{x}(t) = \frac{2}{A+B} \sum_{m=-\infty}^{\infty} \sum_{n=-\infty}^{\infty} \langle x, \psi_{mn} \rangle \psi_{mn}(t). \tag{9.20}$$

If T is chosen just large enough that the samples $W_x(2^m nT, 2^m), m, n \in Z$ contain no redundancy at all (critical sampling), the functions $\psi_{mn}(t), m, n \in Z$ are linearly independent. If (9.19) is also satisfied with $0 < A \leq B < \infty$, the functions $\psi_{mn}(t), m, n \in Z$ form a basis for $L_2(R)$. Then the following relation, which is known as the *biorthogonality condition*, holds:

$$\langle \psi_{mn}, \tilde{\psi}_{lk} \rangle = \delta_{ml}\, \delta_{nk}, \quad m,n,l,k \in \mathbb{Z}. \tag{9.21}$$

Wavelets that satisfy (9.21) are called *biorthogonal wavelets*. As a special case, we have the *orthonormal wavelets*. They are self-reciprocal and satisfy the orthonormality condition:

$$\langle \psi_{mn}, \psi_{lk} \rangle = \delta_{ml}\, \delta_{nk}, \quad m,n,l,k \in \mathbb{Z}. \tag{9.22}$$

Thus, in the orthonormal case, the functions $\psi_{mn}(t)$, $m,n \in Z$ can be used for both analysis and synthesis. Orthonormal bases always have the same frame bounds (tight frame), because, in that case, (9.22) is a special form of Parseval's relation.

9.5 Discrete Wavelet Transform (DWT)

If suitable transformation is applied to a group of selected wavelet, a collection of orthogonal real-valued wavelets will be generated, a representation of the received signal referred to as wavelet expansion. In this case, the properties of the generated wavelets depend on the features of the mother wavelet. Since the newly generated wavelets are a group of orthogonal wavelets, they provide a time-frequency localization of the actual input signal, thereby concentrating the signal energy over a few frequency coefficients. Scaling and translation of the mother wavelet generated. If the scaling factor is a power of two, the wavelet transform technique is referred to as the dyadic-orthonormal wavelet transform. If the chosen mother wavelet has orthonormal properties, there is no redundancy in the discrete wavelet transforms. In addition, this provides the multiresolution algorithm decomposing a signal into scales with different time and frequency resolution. *In this section, the multiresolution analysis and the efficient realization of the* discrete wavelet transform based on multirate filter banks will be addressed. This framework has mainly been developed by Meyer, Mallat and Daubechies for the orthonormal case. Since biorthogonal wavelets formally fit into the same framework, the derivations will be given for the more general biorthogonal case.

9.5.1 Multiresolution Analysis

In the following we assume that the sets

$$\begin{aligned} \psi_{mn}(t) &= 2^{-\frac{m}{2}}\psi(2^{-m}t - n), \\ \tilde{\psi}_{mn}(t) &= 2^{-\frac{m}{2}}\tilde{\psi}(2^{-m}t - n), \end{aligned} \qquad m, n \in \mathbb{Z} \qquad (9.23)$$

are bases for $L_2(\mathrm{R})$ satisfying the biorthogonality condition (9.21). Note that $T = 1$ is chosen in Order to simplify notation. We will mainly consider the representation (9.18) and write it as

$$x(t) = \sum_{m=-\infty}^{\infty} \sum_{n=-\infty}^{\infty} d_m(n)\, \psi_{mn}(t) \qquad (9.24)$$

with

$$d_m(n) = \mathcal{W}_x^{\tilde{\psi}}(2^m n, 2^m) = \left\langle x, \tilde{\psi}_{mn} \right\rangle, \qquad m, n \in \mathbb{Z}. \qquad (9.25)$$

Since a basis consists of linearly independent functions, $L_2(\mathrm{R})$ may be understood as the direct sum of subspaces

$$L_2(\mathbb{R}) = \ldots \oplus W_{-1} \oplus W_0 \oplus W_1 \oplus \ldots \qquad (9.26)$$

With

$$W_m = \text{span}\left\{\psi(2^{-m}t - n),\ n \in \mathbb{Z}\right\}, \quad m \in \mathbb{Z}. \tag{9.27}$$

Each subspace W, covers a certain frequency band. For the subband signals we obtain from (9.24):

$$y_m(t) = \sum_{n=-\infty}^{\infty} d_m(n)\,\psi_{mn}(t), \quad y_m(t) \in W_m. \tag{9.28}$$

Every signal $x(t) \in \mathbf{L}_2(\mathbf{R})$ can be represented as

$$x(t) = \sum_{m=-\infty}^{\infty} y_m(t), \quad y_m(t) \in W_m. \tag{9.29}$$

Now we define the subspaces $V_m, m \in Z$ as the direct sum of $V_m + l$ and $W_m + 1$:

$$V_m = V_{m+1} \oplus W_{m+1}. \tag{9.30}$$

Here we may assume that the subspaces V_m, contain lowpass signals and that the bandwidth of the signals contained in V_m, reduces with increasing m. From (9.27), (9.26), and (9.30) we derive the following properties:

(i) We have a nested sequence of subspaces

$$\ldots \subset V_{m+1} \subset V_m \subset V_{m-1} \subset \ldots \tag{9.31}$$

(ii) Scaling of $x(t)$ by the factor two ($x(t) \to x(2t)$) makes the scaled signal $x(2t)$ an element of the next larger subspace and vice versa:

$$x(t) \in V_m \quad \Leftrightarrow \quad x(2t) \in V_{m-1}. \tag{9.32}$$

(iii) If we form a sequence of functions $x_m(t)$ by projection of $x(t) \in L_2(R)$ onto the subspaces V, this sequence converges towards $x(t)$:

$$\lim_{m \to -\infty} x_m(t) = x(t), \quad x(t) \in L_2(\mathbb{R}), \quad x_m(t) \in V_m. \tag{9.33}$$

Thus, any signal may be approximated with arbitrary precision. Because of the scaling property (9.32) we may assume that the subspaces V_m, are spanned by scaled and time-shifted versions of a single function $\emptyset(t)$

$$V_m = \text{span}\left\{\phi(2^{-m}t - n),\ n \in \mathbb{Z}\right\}. \tag{9.34}$$

Thus, the subband signals $x_m(t) \in V_m$, are expressed as

$$x_m(t) = \sum_{n=-\infty}^{\infty} c_m(n)\,\phi_{mn}(t) \tag{9.35}$$

With

$$\phi_{mn}(t) = 2^{-\frac{m}{2}}\phi(2^{-m}t - n). \tag{9.36}$$

The function $\emptyset(t)$ is called a *scaling function*

Example: Haar Wavelets. The *Haar function* is the simplest example of an orthonormal wavelet:

$$\psi(t) = \begin{cases} 1 & \text{for } 0 \le t < 0.5 \\ -1 & \text{for } 0.5 \le t < 1 \\ 0 & \text{otherwise.} \end{cases}$$

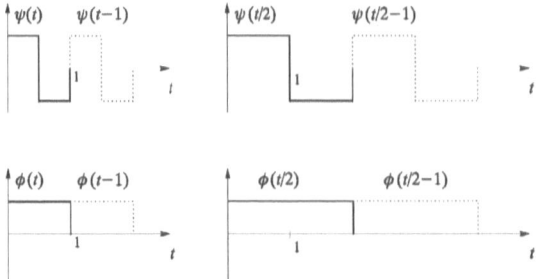

Figure 9.4. Haar wavelet and scaling function

The corresponding scaling function is

$$\phi(t) = \begin{cases} 1, & \text{for } 0 \le t < 1 \\ 0, & \text{otherwise.} \end{cases}$$

The functions $\psi(t-n), n \in Z$ span the subspace W_0, and the functions $\psi(\frac{1}{2}t - n), n \in Z$ span W_1. Furthermore, the functions $\emptyset(t-n), n \in Z$ span V_0 and the functions $\psi(\frac{1}{2}t - n), n \in Z$ span V_1. The orthogonality among the basic functions $\psi(2^{-m}t - n), m, n \in Z$ and the orthogonality of the functions $\psi(2^{-m}t - n), m, n \in Z$ and $\emptyset(2^{-j}t - n), j \ge m$ is obvious, see Figure 8.8.

Example: Shannon Wavelets. The Shannon wavelets are impulse responses of ideal bandpass filters:

$$\psi(t) = \frac{\sin \frac{\pi}{2}t}{\frac{\pi}{2}t} \cos \frac{3\pi}{2}t. \tag{9.37}$$

In the frequency domain this is

$$\Psi(\omega) = \begin{cases} 1 & \text{for } \pi \le |\omega| \le 2\pi, \\ 0 & \text{otherwise.} \end{cases} \tag{9.38}$$

The scaling function that belongs to the Shannon wavelet is the impulse response of the ideal lowpass:

$$\phi(t) = \frac{\sin \pi t}{\pi t}$$

$$\updownarrow$$

$$\Phi(\omega) = \begin{cases} 1 & \text{for } 0 \le |\omega| \le \pi, \\ 0 & \text{otherwise}. \end{cases} \qquad (9.39)$$

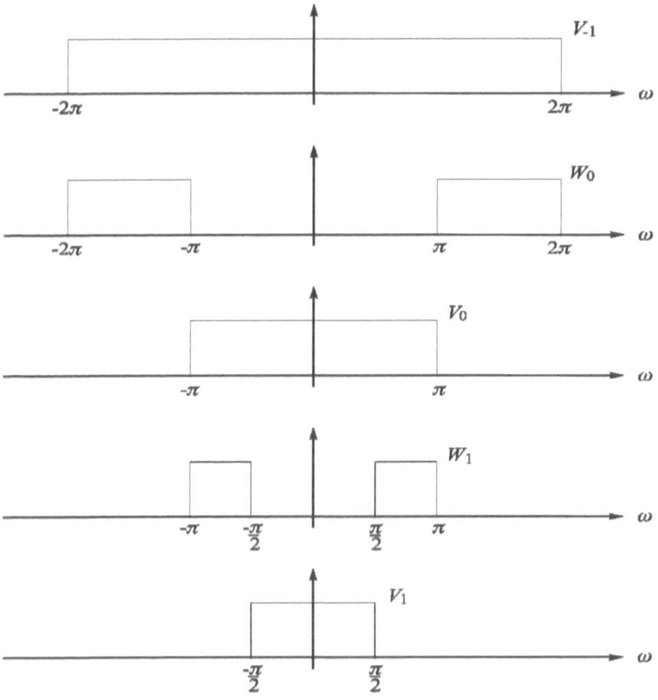

Figure 9.5. Subspaces of Shannon wavelets.

The coefficients $c_m(n), m, n \in Z$ in (9.35) can be understood as the sample values of the ideally lowpass-filtered signal. Figure 8.9 illustrates the decomposition of the signal space.

The Shannon wavelets form an orthonormal basis for $L_2(R)$. The orthogonality between different scales is easily seen, because the spectra do not overlap. For the inner product of translated versions of $\emptyset(t)$ at the same scale, we get

$$\int_{-\infty}^{\infty} \phi(t-m)\phi^*(t-n) = \frac{1}{2\pi}\int_{-\pi}^{\pi} \Phi(\omega)\Phi^*(\omega)e^{-j(m-n)\omega}d\omega$$

$$= \frac{1}{2\pi}\int_{-\pi}^{\pi} e^{-j(m-n)\omega}d\omega$$

$$= \delta_{mn} \qquad (9.40)$$

by using Parseval's relation. The orthogonality of translated wavelets at the same scale is shown using a similar derivation.

A drawback of the Shannon wavelets is their infinite support and the poor time resolution due to the slow decay. On the other hand, the frequency resolution is perfect. For the Haar wavelets, we observed the opposite behavior. They had perfect time, but unsatisfactory frequency resolution.

8.5.2 Wavelet Analysis by Multirate Filtering

Because of $V_0 = V_1 @ W_1$ the functions $\emptyset_{0n}(t) = \emptyset(t-n) \in V_0$, $n \in Z$ can be written as linear combinations of the basis functions for the spaces V_1 and W_1. With the coefficients $h_0(2l-n)$ and $h_1(2l-n), l, n \in Z$ the approach is

$$\phi_{0n}(t) = \sum_{\ell} h_0(2\ell-n)\,\phi_{1\ell}(t) + h_1(2\ell-n)\,\psi_{1\ell}(t).$$
$$(9.41)$$

Equation (9.41) is known as the *decomposition relation*, for which the following notation is used as well:

$$\sqrt{2}\,\phi(2t-n) = \sum_{\ell} h_0(2\ell-n)\,\phi(t-\ell) + h_1(2\ell-n)\,\psi(t-\ell).$$
$$(9.42)$$

We now consider a known sequence $\{c_0(n)\}$, and we substitute (9.41) into (9.35) for $m = 0$. We get

$$\begin{aligned}
x_0(t) &= \sum_n c_0(n)\, \phi_{0n}(t) \\
&= \sum_n c_0(n) \sum_\ell h_0(2\ell - n)\, \phi_{1\ell}(t) + h_1(2\ell - n)\, \psi_{1\ell}(t) \\
&= \sum_\ell \underbrace{\sum_n c_0(n)\, h_0(2\ell - n)}_{c_1(\ell)} \phi_{1\ell}(t) + \sum_\ell \underbrace{\sum_n c_0(n)\, h_1(2\ell - n)}_{d_1(\ell)} \psi_{1\ell}(t) \\
&= x_1(t) + y_1(t),
\end{aligned} \qquad (9.43)$$

where $X_0 \in V_0$, $X_1 \in V_1$, and $Y_1 \in W_1$. This method allows us to compute $\{C_{m+1}(l)\}$ and $\{d_{m+1}(l)\}$ from $\{c_m(n), m, n \in \mathbb{Z}\}$:

$$\left. \begin{aligned}
c_{m+1}(\ell) &= \sum_n c_m(n)\, h_0(2\ell - n) \\
d_{m+1}(\ell) &= \sum_n c_m(n)\, h_1(2\ell - n)
\end{aligned} \right\}, \quad i, \ell \in \mathbb{Z}. \qquad (9.44)$$

We see that the sequences $\{C_{m+1}(l)\}$ and $\{d_{m+1}(l)\}$ occur with half the sampling rate of $\{c_m(n)\}$. Altogether, the decomposition (9.44) is equivalent to a two-channel filter bank analysis with the analysis filters $h_0(n)$ and $h_1(n)$.

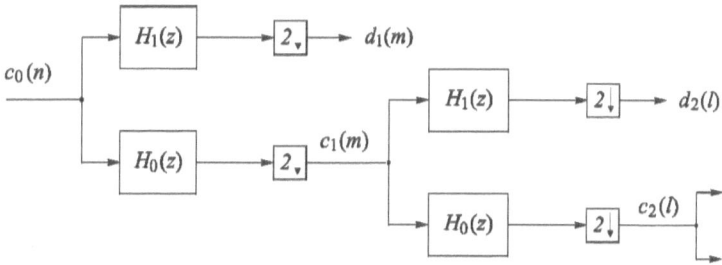

Figure 9.6. Analysis filter bank for computing the DWT.

If we assume that $x_0(t)$ is a sufficiently good approximation of $x(t)$ and if we know the coefficients $c_0(n)$, we can compute the coefficients $C_{m+1}(n)$ & $d_{m+1}(n)$, $m > 0$, and thus the values of the wavelet transform using the discrete-time filter bank depicted in Figure 8.10. This is the most efficient way of computing the DWT of a signal.

8.5.3 Wavelet Synthesis by Multirate Filtering

Let us consider two sequences $g_0(n)$ and $g_1(n)$, which allow us to express the functions $\emptyset_{10}(t) = 2^{-\frac{1}{2}}\emptyset(t/2) \in V_1$ and $\psi_{10}(t) = 2^{-\frac{1}{2}}\psi(t/2) \in W_1$ as linear combinations of $\emptyset_{0n}(t) = \emptyset(t-n) \in V_0, n \in Z$ in the form

$$\phi_{10}(t) = \sum_n g_0(n)\,\phi_{0n}(t),$$
$$\psi_{10}(t) = \sum_n g_1(n)\,\phi_{0n}(t), \tag{8.45}$$

or equivalently as

$$\phi(t) = \sum_n g_0(n)\,\sqrt{2}\,\phi(2t-n),$$
$$\psi(t) = \sum_n g_1(n)\,\sqrt{2}\,\phi(2t-n). \tag{8.46}$$

Equations (9.45) and (9.46), respectively, are referred to as the *two-scale relation*. For time-shifted functions the two-scale relation is

$$\phi_{1\ell}(t) = \sum_n g_0(n-2\ell)\,\phi_{0n}(t),$$
$$\psi_{1\ell}(t) = \sum_n g_1(n-2\ell)\,\phi_{0n}(t). \tag{9.47}$$

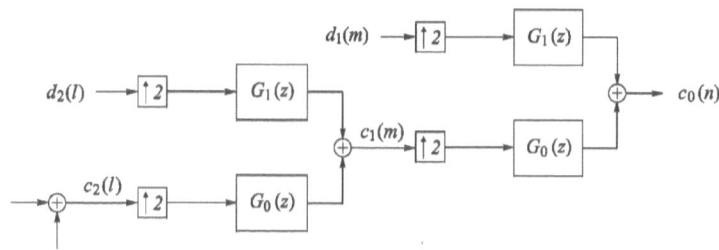

Figure 9.7. Synthesis filter bank

From (9.47), (9.28), (9.35) we can derive

$$\begin{aligned}
x_0(t) &= x_1(t) + y_1(t) \\
&= \sum_\ell c_1(\ell)\, \phi_{1\ell}(t) + \sum_\ell d_1(\ell)\, \psi_{1\ell}(t) \\
&= \sum_\ell c_1(\ell) \sum_n g_0(n-2\ell)\, \phi_{0n}(t) + \sum_\ell d_1(\ell) \sum_n g_1(n-2\ell)\, \phi_{0n}(t) \\
&= \sum_n \left(\sum_\ell c_1(\ell)\, g_0(n-2\ell) + d_1(\ell)\, g_1(n-2\ell) \right) \phi_{0n}(t) \\
&= \sum_n c_0(n)\, \phi_{0n}(t).
\end{aligned} \quad (9.48)$$

The generalization of (9.48) yield

$$c_m(n) = \sum_\ell c_{m+1}(\ell)\, g_0(n-2\ell) + d_{m+1}(\ell)\, g_1(n-2\ell). \quad (9.49)$$

The sequences $g_0(n)$ and $g_1(n)$ may be understood as the impulse responses of discrete-time filters, and (9.49) describes a discrete-time two-channel synthesis filter bank. The filter bank is shown in Figure 9.7.

9.6 Discrete Wavelet Transform for denoising data

The DWT denoising procedure consists of three steps. In the first step, if the length of the data stream is of length of the order of power of two, it is transformed to the wavelet domain.

In the second step, coefficients with either zero magnitude or criterion-based minimized values are selected. In the third or final step, the minimized coefficients are reverted back to the original domain from the wavelet domain to extract the denoised data. DWT-based denoising techniques can be broadly classified into two categories - linear and non-linear. In linear DWT, signal and noise are assumed to be belonging to the smooth and the detailed part of the wavelet domain, where high frequency components are attenuated. While in non-linear DWT, the filter removes the coefficients selected in the second stage with amplitudes less than the threshold. In practicality, non-linear DWT is always preferred over linear DWT, as linear DWT introduces error due to the retention of noise components and loss of signal components owing to wavelet filtering.

Whether linear or non-linear DWT denoising technique is used, performance depends on the choice of the wavelet family and the length of the filter. The traditional way for making this choice is based on visual inspection of the data, for example, Daubechies wavelets are implemented when the data appears smooth in the wavelet domain, while Haar or other wavelets are used when the data appears bursty and discontinuous in the wavelet domain. To overcome the problems with DWT denoising, correlation denoising method is used. Correlation denoising method implements wavelet transformation and filtering in a way such that the correlation between wavelet coefficients of the signal part and the noise part is different at each level. However, correlation denoising in its original form is computationally complex. To reduce computational complexity, wavelet threshold denoising method is used. The method is simple to calculate, and the noise can be suppressed to a large extent. At the same time, singular information of the original signal can be preferred well, so it is a simple and effective method. A brief overview of what happens when DWT is applied for denoising is demonstrated in figure 9.8.

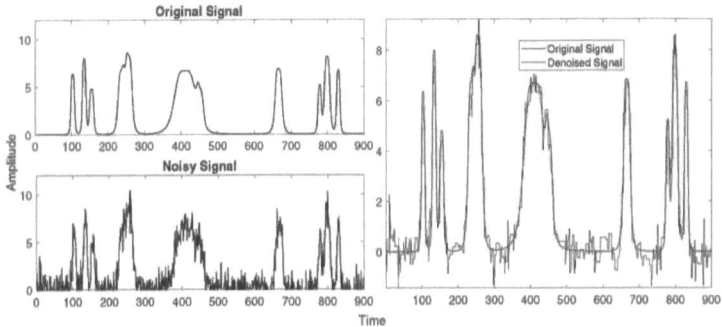

Figure 9.8. Denoising with DWT.

The four major components of the DWT denoising technique are: wavelet-type selection, threshold selection, threshold function selection and threshold application to the wavelet coefficients.

Wavelet Selection, there is a wide variety of wavelets that can be used for denoising. Selecting the optimum one depends on the selection of the matching wavelet filter. Out of different wavelet transform based denoising methods, only minimum description length (MDL) method has the flexibility of choosing the filter type.

9.7 Signal denoising for IoT networks

The huge amount of sensor data generated in an IoT network are used to take decisions on a certain observation/ phenomenon based on real-time processing. The decision-making procedure often involves detecting the signal energy level transmitted from the sensors. If the received energy level is higher than a predefined threshold, the target is detected to be present phenomenon and vice-versa. However, the sensor data gets crippled with noise contributed by the wireless environment and the internal electronics of the sensors, on its way to the data center for processing. The WPT method will be the best option in this case for denoising the sensor data, where the original signal coefficients are preserved while removing the noise within the signal. The WPT method can decompose a signal in both scale and wavelet space thereby revealing more details about both the sensor signals and the crippling noise. If energy correlation analysis is used in conjunction with WPT, signal energy from the sensor data can be analyzed and noise can be eliminated by zooming into the signal characteristics at different time scales. Advantages of WPT over WT is evident in Figure 9.9. Hence, in this section, a universal framework is presented for denoising sensor signals in IoT networks. The framework is based on energy correlation analysis and combines the processes of WP decomposition, coefficient modification and WP reconstruction. The functional block diagram for this framework is presented in figure 9.9.

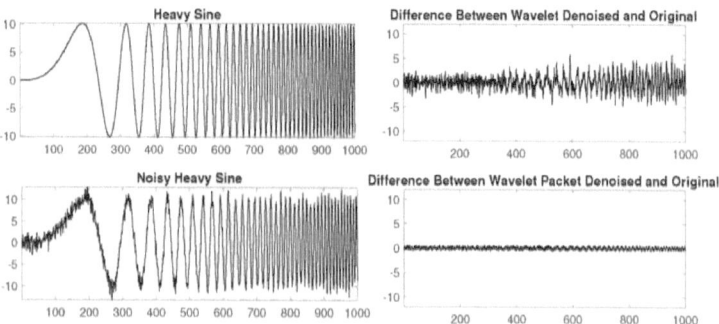

Figure 9.9. Comparative performance of WPT and WT.

9.8 Multiresolution Signal Analysis

In this subsection we show an example of multiresolution analysis for a simple transient signal. Transient signals in the power system are nonstationary time-varying voltage and current that can occur because of changes in the electrical configuration and in industrial and residential loads, and of a variety of disturbances on transmission lines, including capacitor switching, lightning strikes and short-circuits. The waveform data of the transient signals are captured by digital transient recorders. Analysis and classification of the power system disturbance can help to provide more stability and efficiency in power delivery by switching transmission lines to supply additional current or switching capacitor banks to balance inductive loads and help to prevent system failures.

The power system transient signals contain a range of frequencies from a few hertz to impulse components with microsecond rise times.

The normal 60 Hz sinusoidal voltage and current waveforms are interrupted or superimposed with impulses, oscillations, and reflected waves. An experienced power engineer can visually analyze the waveform data in order to determine the type of system disturbance. However, the Fourier analysis with its global operation nature is not as appropriate for the transient signals as the time-scale joint representation provided by the wavelet transform.

9.9 Multiresolution Wavelet Decomposition of Transient Signal

The wavelet transform provides a decomposition of power system transient signals into meaningful components in multiple frequency bands, and the digital wavelet transform is computationally efficient. Figure 9.10 shows the wavelet components in the multiple frequency bands. At the top is the input voltage transient signal. There is a disturbance of a capacitor bank switching on a three-phase transmission line. Below the first line are the wavelet components as a function of the scale and time shift. The scales of the discrete wavelets increase by a factor of two successively from scale 1 to scale 64, corresponding to the dyadic frequency bands. The vertical axis in each discrete scale is the normalized magnitude of the signal component in voltage. The three impulses in high frequency band scale 1 correspond to the successive closing of each phase of the three-phase capacitor bank. scale 2 and scale 4 are the bands of system response frequencies. Scale 4 contains the most energy from the resonant frequency caused by the addition of a capacitor bank to a primarily inductive circuit. The times of occurrence of all those components can be determined on the time axis. Scale 64 contains the basic signal of continuous 60 Hz.

The wavelet analysis decomposes the power system transient into the meaningful components, whose modulus maxima then can be used for further classification. The nonorthogonal multiresolution analysis wavelets with finite impulse response (FIR) quadratic spline wavelet filters were used in this example of application.

One problem in this application and many other applications with the dyadic wavelet transform is the lack of shift invariance. The dyadic wavelet transform is not shift invariant. In the wavelet decomposition the analysis low-pass and high-pass filters are double shifted by two. If the input signal is shifted by one sampling interval distance, the output of the dyadic wavelet transform is not simply shifted by the same distance, but the values of the wavelet coefficients would be changed dramatically. This aliasing error is caused by the down-sampled factor of two in the multiresolution signal analysis. This is a disadvantage of the dyadic wavelet transform because many applications such as real-time signal analysis and pattern recognition require shift invariant wavelet transform. In the above example of application, the orthonormal quadrature mirror filters have been found sensitive to translations of the input. Hence, nonorthonormal quadratic spline wavelets have been used.

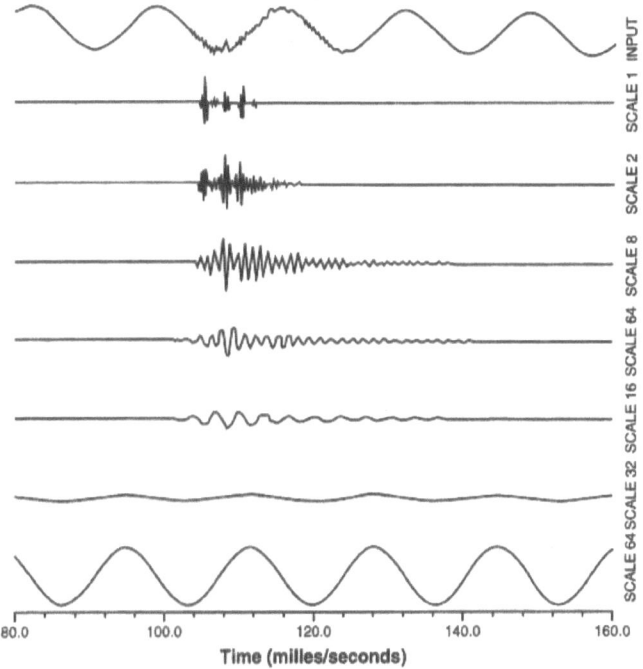

Figure 9.10. Multiresolution wavelet analysis of a transient signal in the electrical power system

9.10 Signal Detection

The detection of weak signals embedded in a stronger stationary stochastic process, such as the detection of radar and sonar signals in zero-mean Gaussian white noise, is a well-studied problem. If the shape of the expected signal is known, the correlation and the matched filter provide optimum solution in terms of the signal-to-noise ratio in the output correlation.

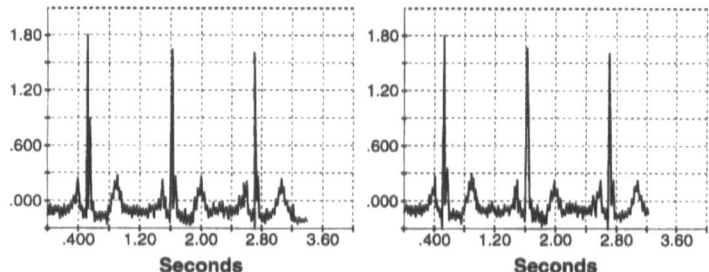

Figure 9.11. (Left) Normal electrocardiogram, (right) electrocardiogram with VLP abnormality

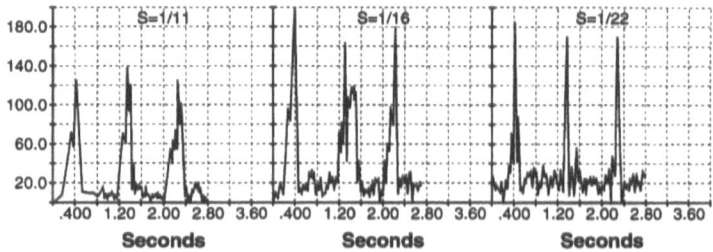

Figure 9.12. Wavelet transform of the abnormal electrocardiogram for scale factor $s = 11, 16, 22$. The bulge to the right of the second QRS peak for $s = 1/16$ indicates the presence of the VLP

In the detection of speech or biomedical signals, the exact shape of the signal is unknown. The Fourier spectrum analysis could be effective for those applications, only when the expected signal has spectral features that clearly distinguishes it from the noise. The effectiveness of the Fourier spectrum analysis is generally proportional to the ratio of the signal to noise energy. For short-time, low-energy transients, the change in the Fourier spectrum is not easily detected. Such transient signals can be detected by the wavelet transform. An example of an electrocardiogram signal detection follows. Figure 9.11 shows the clinical electrocardiogram with normal QRS peaks and an abnormality called ventricular late potentials (VLP) right after the second QRS peak. The amplitude of the VLP signal is about

5% of the QRS peaks. Its duration was about 0.1 second, or a little less than 10% of the pulse period. The VLPs are weak signals, swamped by noise, and they occur somewhat randomly. Figure 9.12. shows the magnitude of continuous wavelet transform with the cos-Gaussian wavelets of scale $s = 1/11$, $1/16$ and $1/22$. The peak after the second QRS spike observed for $s = 1/16$ is very noticeable and gives a clear indication of the presence of the VLP.

Problems

9.1 Find (by hand) what the signals $z_d[n]$ and $w_d[n]$ would be for the _lter bank in Figure Problem 9.1. Let $x = [8, 4, 0, 6, 3, 7, 2, 9]$, $a = \frac{1}{2}$ and $b = \frac{1}{2}$. Be sure to show your work.

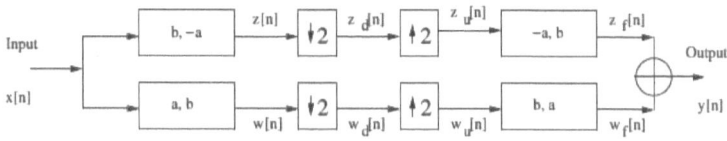

Figure Problem 9.1

9.2 What is multiresolution (i.e., a wavelet transform having more than one octave)? Demonstrate this idea with a figure.

9.3 If the down-samplers keep only one of every four values, what effect would this have on the filter bank? What if the filter bank structure were modified to have four channels?

9.4. Write a function to return the one octave, Daubechies four-coefficient wavelet transform for a given signal. Include low- and highpass outputs.

9.5. Given the input signal $x[n] = 2\cos(2\pi 100n/300) + \cos(\frac{2\pi 110n}{300} - \pi)$ for $n = 0..255$, write the commands to find the DWT for 3 octaves. Compare your results with those of the dwt function. Plot the original function, as well as the approximate signals.

9.6. For the Haar transform, use values $a = b = \frac{1}{2}$ (no down sampling), find signals z, w, and y, given an input of $x = \{6, 1, 3, 7, 2, 5, 8, 10\}$.

9.7. Suppose you have a 3-octave DWT. Draw the analysis structure in terms of filters and down-samplers.

9.8. For a four-octave DWT, suppose the input has 1024 samples. How long would the detail outputs be? How long would the approximate outputs be? What if down/up-sampling were not used? For simplicity, you can assume that the filter's outputs are the same lengths as their inputs.

9.9. For an input of 1024 samples, how many octaves of the DWT could we have before the approximation becomes a single number? For simplicity, you can assume that the filter's outputs are the same lengths as their inputs.

Chapter Ten

Adaptive Signal Processing

Learning Outcomes of this Chapter

After successful completion of this chapter students will be able to:

1. use basic probability theory to model random signals in terms of Random Processes.
2. understand and derive the Wiener filter for signals with known second order statistics.
3. formulate the Wiener filter as a constrained optimization problem.
4. use and understand the LMS algorithm for iteratively estimating the Wiener filter weights.
5. derive and apply the RLS algorithm for iteratively estimating the Wiener filter weights.

10.1 Introduction

Adaptive signal processing is the design of adaptive systems for signal-processing applications. In signal measurement systems the information-bearing signal is often contaminated by noise from its surrounding environment. The noisy observation, $y(n)$, can be modelled as $y(n) = x(n) + n(n)$ where $x(n)$ and $n(n)$ are the signal and the noise, and m is the discrete-time index. In some situations, for example when using a mobile telephone in a moving car, or when using a radio communication device in an aircraft cockpit, it may be possible to measure and estimate the instantaneous amplitude of the ambient noise using a directional microphone. The signal, x(m), may then be recovered by subtraction of an estimate of the noise from the noisy signal.

Figure 10.1 shows a two-input adaptive noise cancellation system for enhancement of noisy speech. In this system a directional microphone takes as input the noisy signal $x(n) + n(n)$, and a second directional microphone, positioned some distance away, measures the noise $\alpha n(n + \zeta)$. The attenuation factor α and the time delay ζ provide a rather over-simplified model of the effects of propagation of the noise to different positions in the space where the microphones are placed. The noise from the second microphone is processed by an adaptive digital filter to make it equal to the noise contaminating the speech signal, and then subtracted from the noisy signal to cancel out the noise.

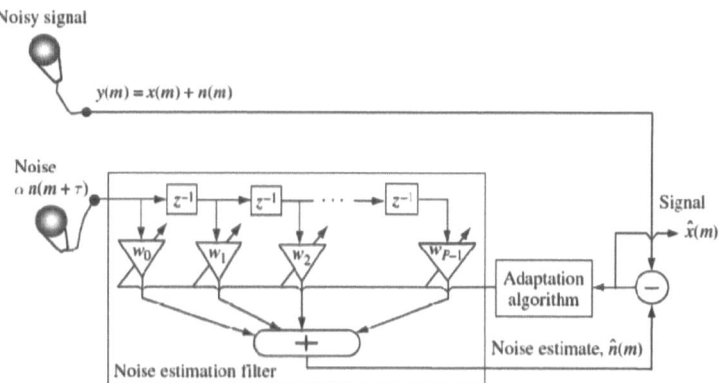

Figure 10.1. A two-input adaptive noise cancellation system

The adaptive noise canceller is more effective in cancelling out the low-frequency part of the noise, but generally suffers from the non-stationary character of the signals, and from the over-simplified assumption that a linear filter can model the diffusion and propagation of the noise sound in the space.

10.2 Adaptive Noise Cancellation

One of the most important applications of adaptive filtering is adaptive noise cancellation (ANC). The background noise is an important handicap and challenging task. If it is joined with other distortions, it can seriously damage the service quality. In all applications that require at least one microphone, the signal of interest is usually contaminated by background noise and reverberation. As a result, the microphone signal has to be "cleaned" with digital signal processing tools before it is played out, transmitted or stored. So, it is important to cancel the noise which may combine the signal to obtain a good quality signal, this may be achieved using adaptive Noise Cancellation which improves the Signal-to-Noise Ratio at the received noisy signal.

Adaptive noise cancellation is an alternative technique of estimating signals corrupted by additive noise or interference. It has advantage that, with no prior knowledge of signal or noise, levels of noise reject ion are obtainable that would be difficult to achieve by other signal processing methods of noise removing. The principle of adaptive noise cancellation is to obtain an estimate of the noise signal and subtract it from the corrupted signal.

As shown in the figure, an Adaptive Noise Canceller (ANC) has two inputs – primary and reference. The primary input receives a signal s from the signal source.

that is corrupted by the presence of noise n uncorrelated with the signal. The reference input receives a noise n_0 uncorrelated with the signal but correlated in some way with the noise n.

The noise no passes through a filter to produce an output n^ that is a close estimate of primary input noise. This noise estimate is subtracted from the corrupted signal to produce an estimate of the signal at s^, the ANC system output.

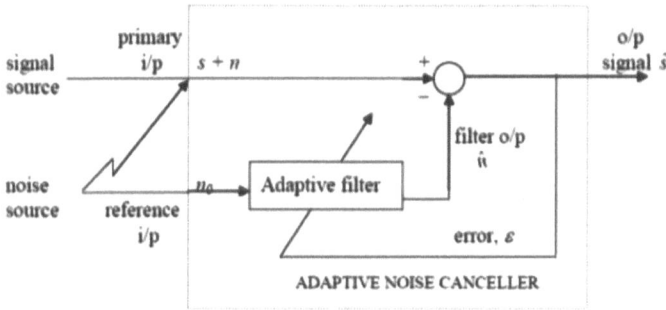

Figure 10.2 Adaptive Noise Canceller

The adaptive noise cancellation technique uses adaptive filters for signal processing. The adaptive filter constitutes an important part in statistical signal processing. Whenever there is a requirement to process signals that result from operation in an environment of unknown statistics, the use of an adaptive filter offers an attractive solution to the problem as it usually provides a significant improvement in performance over the use of a fixed filter designed by conventional methods. Furthermore, the use of adaptive filters provides new signal-processing capabilities that would not be possible otherwise.

We thus find that adaptive filters are successfully applied in such diverse fields as communications, control, radar, sonar, seismology, and biomedical engineering.

An adaptive filter is very generally defined as a filter whose characteristics can be modified to achieve some end or objective and is usually assumed to accomplish this modification or adaptation automatically.

10.3 Adaptive Filtering Algorithms

Adaptive filtering is the process which is required for echo canceling in different applications. Adaptive filter is such type of filter whose characteristics can be changed for achieving optimal desired output. An adaptive filter can change its parameters to minimize the error signal by using adaptive algorithms. The error is the difference between the desired signal and the output signal of the filter. Therefore, a digital filter that automatically adjusts its coefficients to adapt input signal via an adaptive algorithm. Characteristics of adaptive filters: They can automatically adapt (self-optimize) in the face of changing environments and changing system requirements. They can be trained to perform specific filtering and decision-making tasks according to some updating equations (training rules). An adaptive filter is defined by four aspects:

1. the signals being processed by the filter.
2. the structure that defines how the output signal of the filter is computed from its input signal.
3. the parameters within this structure that can be iteratively changed to alter the filter's input-output relationship 4. the adaptive algorithm that describes how the parameters are adjusted from one time instant to the next.

Figure 10.3 shows a block diagram in which a sample from a digital input signal $x(n)$ is fed into a device, called an adaptive filter, that computes a corresponding output signal sample $y(n)$ at time n. For the moment, the structure of the adaptive filter is not important, except for the fact that it contains adjustable parameters whose values affect how y(n) is computed. The output signal is compared to a second signal d(n), called the desired response signal, by subtracting the two samples at time n. This difference signal, given by:

$$e(n) = d(n) - y(n) \qquad (10.1)$$

is known as the error signal. The error signal is fed into a procedure which alters or adapts the parameters of the filter from time n to time (n + 1) in a well-defined manner. This process of adaptation is represented by the oblique arrow that pierces the adaptive filter block in the figure. As the time index *n* is incremented, it is hoped that the output of the adaptive filter becomes a better and better match to the desired response signal through this adaptation process, such that the magnitude of *e*(*n*) decreases over time. In this context, what is meant by "better" is specified by the form of the adaptive algorithm used to adjust the parameters of the adaptive filter. In the adaptive filtering task, adaptation refers to the method by which the parameters of the system are changed from time index *n* to time index (*n* + 1). The number and types of parameters within this system depend on the computational structure chosen for the system. We now discuss different filter structures that have been proven useful for adaptive filtering tasks.

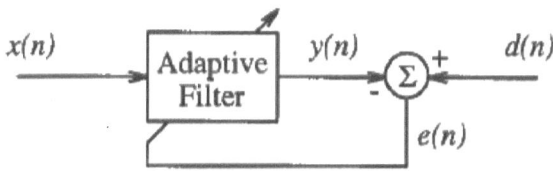

Figure 10.3 The general adaptive filtering problem

In general, any system with a finite number of parameters that affect how *y*(*n*) is computed from *x*(*n*) could be used for the adaptive filter in Figure 10.3. Define the *parameter* or *coefficient vector w*(*n*) as

$$w(n) = [w_o(n), w_1(n), w_2(n), \ldots \ldots \ldots \ldots \ldots, w_{L-1}(n)]^T \quad (10.2)$$

where $\{w_i(n)\}$, $0 \leq i \leq L-1$ are the L parameters of the system at time n. With this definition, we could define a general input-output relationship for the adaptive filter as :

$$y(n) = f(W(n), y(n-1), y(n-2),\ldots y(n-N), x(n), x(n-1), \ldots, x(n-M+1)) \quad (10.3)$$

where $f(.)$ represents any well-defined linear or nonlinear function and M and N are positive integers. Implicit in this definition is the fact that the filter is *causal*, such that future values of $x(n)$ are not needed to compute $y(n)$. While noncausal filters can be handled in practice by suitably buffering or storing the input signal samples, we do not consider this possibility.

Although equation 10.3 is the most general description of an adaptive filter structure, we are interested in determining the best *linear relationship* between the input and desired response signals for many problems. This relationship typically takes the form of a *finite-impulse-response* (FIR) or *infinite impulse-response* (IIR) filter. Figure 10.3 shows the structure of a direct-form FIR filter, also known as a *tapped-delay-line* or *transversal filter*, where z^{-1} denotes the unit delay element and each $w_i(n)$ is a multiplicative gain within the system. In this case, the parameters in $W(n)$ correspond to the impulse response values of the filter at time n. We can write the output signal $y(n)$ as:

$$\begin{aligned} y(n) &= \sum_{i=0}^{L-1} w_i(n) x(n-i) \\ &= \mathbf{W}^T(n)\mathbf{X}(n), \end{aligned} \quad (10.4)$$

where $X(n) = [x(n)\ x(n-1)\ \ldots\ldots\ldots\ x(n-L+1)]^T$ denotes the *input signal vector* and $.^T$ denotes vector transpose. Note that this system requires L multiplies and $L-1$ adds to implement, and these computations are easily performed by a processor or circuit so long as L is not too large and the sampling period for the signals is not too short. It also requires a total of $2L$ memory locations to store the L input signal samples and the L coefficient values, respectively.

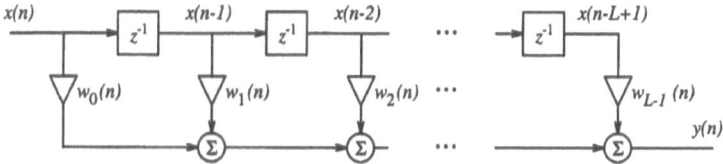

Figure 10.3. Structure of an FIR filter.

10.3.1 Least Mean Square (LMS) Algorithm

It is a class of adaptive filter used to mimic a desired filter by finding the filter coefficients that relate to producing the least mean squares of the error signal (difference between the desired and the actual signal). The simplicity of the Least Mean Square (LMS) algorithm and ease of implementation makes it the best choice for many real-time systems. It is a stochastic gradient descent method in that the filter is only adapted based on the error at the current time. It was invented in 1960 by Stanford University professor Bernard Widrow and his first Ph.D. student, Ted Hoff. The simplicity of the Least Mean Square (LMS) algorithm and ease of implementation makes it the best choice for many real-time systems.

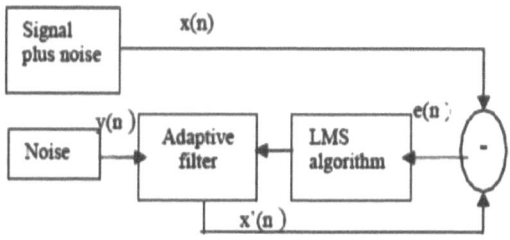

Figure 10.4 Adaptive LMS Filter

The implementation steps for this algorithm can be stated as;

1. Define the desired response and set each coefficient weight to zero.

$$w(n) = 0, \quad n = 1,2,3, \ldots, N \quad (10.5)$$

For each sampling instant (n) carry out steps (2) to (4):

2. Move all the samples in the input array one position to the right, now load the current data sample n into the first position in the array. Calculate the output of the adaptive filter by multiplying each element in the array of filter coefficients by the corresponding element in the input array and all the results are summed to give the output corresponding to that data that was earlier loaded into the input array.

$$y(n) = \sum_{n\ 0}^{N\ 1} w(n) x(n) \quad (10.6)$$

3. Before the filter coefficients can be updated the error must be calculated, simply find the difference between the desired response and the output of the adaptive filter.

$$e(n) = y(n) - d(n) \quad (10.7)$$

4. To update the filter coefficients multiply the error by μ, the learning rate parameter and then multiply the result by the filter input and add this result to the values of the previous filter coefficients.

$$\vec{w}(n+1) = \vec{w}(n) + \mu \cdot e(n) \cdot \vec{x}(n) \quad (10.8)$$

where

> μ: is the step size of the adaptive filter
> $\vec{w}(n)$ Is the filter coefficients vector
> $\vec{x}(n)$ Is the filter input vector

Then LMS algorithms calculate the cost function J(n) by using the following equation:

$$J(n) = e^2(n) \qquad (10.9)$$

Where e^2 (n) is the square of the error signal at time *n*

10.3.2 The Recursive Least Squares (RLS) Algorithm

The LMS algorithm has many advantages (due to its computational simplicity), but its convergence rate is slow. The LMS algorithm has only one adjustable parameter that affects convergence rate, the step-size parameter μ, which has a limited range of adjustment in order to insure stability For faster rates of convergence, more complex algorithms with additional parameters must be used. The RLS algorithm uses a least-squares method to estimate correlation directly from the input data. The LMS algorithm uses the statistical mean-squared-error method, which is slower. The standard RLS algorithm performs the following operations to update the coefficients of an adaptive filter.

1. Calculates the output signal *y(n)* of the adaptive filter.
2. Calculates the error signal *e(n)* by using the following equation:

$$e(n) = d(n) - y(n).$$

3. Updates the filter coefficients by using the following equation:

$$\vec{w}(n+1) = \vec{w}(n) + e(n) \cdot \vec{K}(n) \qquad (10.10)$$

Where $\vec{w}(n)$ is the filter coefficients vector and $\vec{K}(n)$ is the gain vector. $\vec{K}(n)$ is defined by the following equation:

$$\vec{K}(n) = \frac{P(n) \cdot \vec{u}(n)}{\lambda + \vec{u}^T(n) \cdot P(n) \cdot \vec{u}(n)} \qquad (10.11)$$

Where λ is the forgetting factor and P(n) is the inverse correlation matrix of the input signal. P(n) has the following initial value P(0):

$$P(0) = \begin{bmatrix} \delta^{-1} & & 0 \\ & \delta^{-1} & \\ & & \ddots \\ 0 & & \delta^{-1} \end{bmatrix}$$

Where δ is the regularization factor. The standard RLS algorithm uses the following equation to update this inverse correlation matrix.

$$P(n+1) = \lambda^{-1} P(n) - \lambda^{-1} \vec{K}(n) \cdot \vec{u}^T(n) \cdot P(n) \qquad (10.12)$$

RLS algorithms calculate J(n) by using the following equation

$$J(n) = \frac{1}{N} \sum_{i=0}^{N-1} \lambda^i \, e^2(n-i) \qquad (10.13)$$

Where N is the filter length and λ is the forgetting factor.

This algorithm calculates not only the instantaneous value $e^2(n)$ but also the past values, such as $e^2(n-1)$, $e^2(n-2)$... $e^2(n-N+1)$. The value range of the forgetting factor is (0, 1]. When the forgetting factor is less than 1, this factor specifies that this algorithm places a larger weight on the current value and a smaller weight on the past values. The resulting $E[e^2(n)]$ of the RLS algorithms is more accurate than that of the LMS algorithms. The LMS algorithms require fewer computational resources and memory than the RLS algorithms. However, the eigenvalue spread of the input correlation matrix, or the correlation matrix of the input signal, might affect the convergence speed of the resulting adaptive filter. The convergence speed of the RLS algorithms is much faster than that of the LMS algorithms. However, the RLS algorithms require more computational resources than the LMS algorithms.

10.3.3 Wiener Filtering

Wiener filters play a central role in a wide range of applications such as linear prediction, echo cancellation, signal restoration, channel equalization and system identification. Wiener filter theory provides a convenient method of mathematically analysing statistical noise cancelling problems. The Wiener filter is a popular technique that has been used in many signal enhancement methods. The basic principle of the Wiener filter is to obtain estimate of speech signal from that corrupted by additive noise. This estimate is obtained by minimizing the Mean Square Error (MSE) between the desired signal *s(n)* and the estimated signal ^s(n). It is based on a statistical approach. The Wiener filter weights noisy signal spectrum according to SNR at different frequencies.

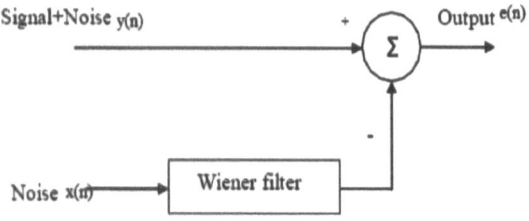

Figure 10.5 Basic of Wiener Filter

Typical filters are designed for a desired frequency response. However, the design of wiener filter takes a different approach. One is assumed to have knowledge of the spectral properties of the original signal and the noise, and one seeks the linear time invariant filter whose output would come as close to the original signal as possible.

1. For this transfer function of wiener filter is used in frequency domain which is expressed as follows:

$$H(\omega) = \frac{P_s(\omega)}{P_s(\omega) + P_d(\omega)}$$

Where, $Ps(\omega)$ and $Pd(\omega)$ are power spectral densities of clean and noisy signals respectively.

2. In wiener filter, the signal and noise is assumed uncorrelated and stationary, and the SNR is given by:

$$SNR = \frac{Ps(\omega)}{Pd(\omega)}$$

3. Using this definition of SNR, the transfer function of Wiener filter can be given

$$H(\omega) = \left[1 + \frac{1}{SNR}\right]^{-1}$$

From the above definition of transfer function, it can be interpreted that the Wiener filter has fixed frequency response at all frequencies and needs an estimation of the power spectral density of clean signal and noise prior to filtering.

10.3.3.1 Adaptive Wiener Filter

This section presents an adaptive implementation of the Wiener filter which benefits from the varying local statistics of the speech signal. The designed adaptive wiener filter depends on the adaptation of the filter transfer function from sample to sample based on the speech signal statistics (mean & variance). A block diagram of the approach is as shown in figure below.

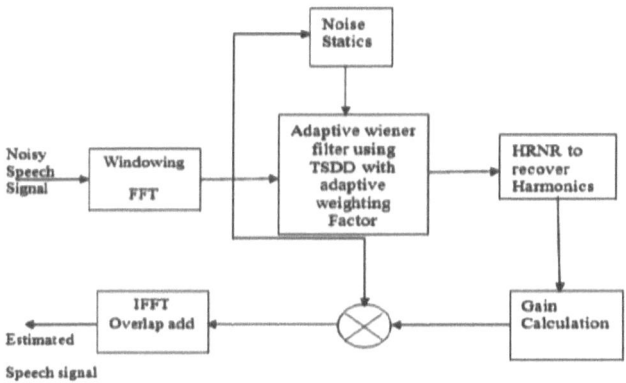

Figure 10.6 Block Diagram of Adaptive Wiener Filter

As we seen above main aim of wiener filter is to find out the signal estimate. This signal estimate is calculated by multiplying spectral gain with noisy speech spectrum. This spectral gain depends upon the *priori* SNR. This priori SNR follows the shape of posterior SNR but with delay of one frame. In practical implementations of speech enhancement systems, the power spectrum density of the speech and the noise are unknown as only the noisy speech is available. Then, both the *instantaneous* SNR and the *a priori* SNR have to be estimated.

This *instantaneous* SNR and the *a priori* SNR can be estimated using decision directed approach. The behaviour of the estimator of the *a priori* SNR controlled by the parameter α. The multiplicative gain function is obtained by multiplying functions of priori SNR with instantaneous SNR. The speech signal spectrum is calculated by multiplying this multiplicative gain with noisy speech spectrum. The Multiplicative gain used here is of wiener transfer function.

10.4 Applications of Adaptive Filters

Perhaps the most important driving forces behind the developments in adaptive filters throughout their history has been the wide range of applications in which such systems can be used. We now discuss the forms of these applications in terms of more-general problem classes that describe the assumed relationship between $d(n)$ and $x(n)$. Our discussion illustrates the key issues in selecting an adaptive filter for a particular task.

10.4.1 System Identification

Consider Figure 10.7, which shows the general problem of *system identification*. In this diagram, the system enclosed by dashed lines is a "black box," meaning that the quantities inside are not observable from the outside. Inside this box is (1) an unknown system which represents a general input output relationship and (2) the signal $\eta(n)$, called the *observation noise signal* because it corrupts the observations of the signal at the output of the unknown system.

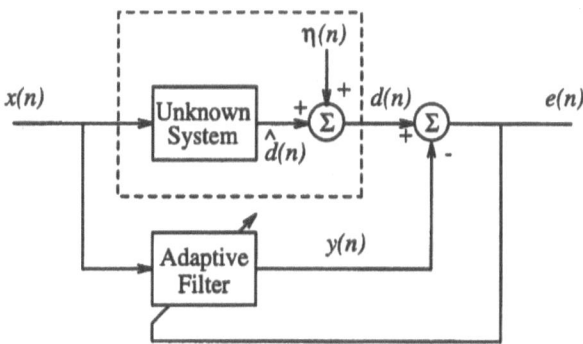

Figure 10.7: System identification

Let $\hat{d}(n)$ represent the output of the unknown system with $x(n)$ as its input. Then, the desired response signal in this model is

$$\hat{d}(n) = d(n) + \eta(n) \tag{10.10}$$

Here, the task of the adaptive filter is to accurately represent the signal $\hat{d}(n)$ at its output. If $y(n) = \hat{d}(n)$, then the adaptive filter has accurately modeled or identified the portion of the unknown system that is driven by $x(n)$.

Since the model typically chosen for the adaptive filter is a linear filter, the practical goal of the adaptive filter is to determine the best linear model that describes the input-output relationship of the unknown system. Such a procedure makes the most sense when the unknown system is also a linear model of the same structure as the adaptive filter, as it is possible that $y(n) = \hat{d}(n)$ for some set of adaptive filter parameters. For ease of discussion, let the unknown system and the adaptive filter both be FIR filters, such that

$$d(n) = W_{opt}^T(n)X(n) + \eta(n) \qquad (10.11)$$

where W_{opt}^T is an optimum set of filter coefficients for the unknown system at time n. In this problem formulation, the ideal adaptation procedure would adjust $W(n)$ such that $W(n) = W_{opt}^T$ as $n \to \infty$. In practice, the adaptive filter can only adjust $W(n)$ such that $y(n)$ closely approximates $\hat{d}(n)$ over time.

The system identification task is at the heart of numerous adaptive filtering applications. We list several of these applications here.

10.4.2 Channel Identification

In communication systems, useful information is transmitted from one point to another across a medium such as an electrical wire, an optical fiber, or a wireless radio link. Nonidealities of the transmission medium or *channel* distort the fidelity of the transmitted signals, making the deciphering of the received information difficult. In cases where the effects of the distortion can be modeled as a linear filter, the resulting "smearing" of the transmitted symbols is known as *inter-symbol interference* (ISI). In such cases, an adaptive filter can be used to model the effects of the channel ISI for purposes of deciphering the received information in an optimal manner. In this problem scenario, the transmitter sends to the receiver a sample sequence x(n) that is known to both the transmitter and receiver. The receiver then attempts to model the received signal d(n) using an adaptive filter whose input is the known transmitted sequence *x.n/:* After a suitable period of adaptation, the parameters of the adaptive filter in W(n) are fixed and then used in a procedure to decode future signals transmitted across the channel.

Channel identification is typically employed when the fidelity of the transmitted channel is severely compromised or when simpler techniques for sequence detection cannot be used.

10.4.3 Plant Identification

In many control tasks, knowledge of the transfer function of a linear plant is required by the physical controller so that a suitable control signal can be calculated and applied. In such cases, we can characterize the transfer function of the plant by exciting it with a known signal x(n) and then attempting to match the output of the plant d(n) with a linear adaptive filter. After a suitable period of adaptation, the system has been adequately modeled, and the resulting adaptive filter coefficients in W(n) can be used in a control scheme to enable the overall closed-loop system to behave in the desired manner. In certain scenarios, continuous updates of the plant transfer function estimate provided by W(n) are needed to allow the controller to function properly.

10.4.4 Echo Cancellation for Long-Distance Transmission

In voice communication across telephone networks, the existence of junction boxes called *hybrids* near either end of the network link hampers the ability of the system to cleanly transmit voice signals. Each hybrid allows voices that are transmitted via separate lines or channels across a long-distance network to be carried locally on a single telephone line, thus lowering the wiring costs of the local network. However, when small impedance mismatches between the long distance lines and the hybrid junctions occur, these hybrids can reflect the transmitted signals back to their sources, and the long transmission times of the long-distance network—about 0.3 s for a trans-oceanic call via a satellite link—turn these reflections into a noticeable echo that makes the understanding of conversation difficult for both callers. The traditional solution to this problem prior to the advent of the adaptive filtering solution was to introduce significant loss into the long-distance network so that echoes would decay to an acceptable level before they became perceptible to the callers.

Unfortunately, this solution also reduces the transmission quality of the telephone link and makes the task of connecting long distance calls more difficult. An adaptive filter can be used to cancel the echoes caused by the hybrids in this situation. Adaptive filters are employed at each of the two hybrids within the network. The input $x.n/$ to each adaptive filter is the speech signal being received prior to the hybrid junction, and the desired response signal $d.n/$ is the signal being sent out from the hybrid across the long-distance connection. The adaptive filter attempts to model the transmission characteristics of the hybrid junction as well as any echoes that appear across the long-distance portion of the network. When the system is properly designed, the error signal $e.n/$ consists almost totally of the local talker's speech signal, which is then transmitted over the network.

10.4.5 Acoustic Echo Cancellation

A related problem to echo cancellation for telephone transmission systems is that of acoustic echo cancellation for conference-style speakerphones. When using a speakerphone, a caller would like to turn up the amplifier gains of both the microphone and the audio loudspeaker to transmit and hear the voice signals more clearly. However, the feedback path from the device's loudspeaker to its input microphone causes a distinctive *howling* sound if these gains are too high.

In this case, the culprit is the room's response to the voice signal being broadcast by the speaker; in effect, the room acts as an extremely poor hybrid junction, in analogy with the echo cancellation task discussed previously. A simple solution to this problem is to only allow one person to speak at a time, a form of operation called *half-duplex transmission*. However, studies have indicated that half-duplex transmission causes problems with normal conversations, as people typically overlap their phrases with others when conversing.

To maintain *full-duplex transmission,* an acoustic echo canceller is employed in the speakerphone to model the acoustic transmission path from the speaker to the microphone. The input signal $x.n/$ to the acoustic echo canceller is the signal being sent to the speaker, and the desired response signal $d.n/$ is measured at the microphone on the device. Adaptation of the system occurs continually throughout a telephone call to model any physical changes in the room acoustics. Such devices are readily available in the marketplace today. In addition, similar technology can and is used to remove the echo that occurs through the combined radio/room/telephone transmission path when one places a call to a radio or television talk show.

10.4.6 Adaptive Noise Cancelling

When collecting measurements of certain signals or processes, physical constraints often limit our ability to cleanly measure the quantities of interest. Typically, a signal of interest is linearly mixed with other extraneous noises in the measurement process, and these extraneous noises introduce unacceptable errors in the measurements. However, if a linearly related *reference* version of any one of the extraneous noises can be cleanly sensed at some other physical location in the system, an adaptive filter can be used to determine the relationship between the noise reference $x(n)$ and the component of this noise that is contained in the measured signal $d(n)$. After adaptively subtracting out this component, what remains in $e.n/$ is the signal of interest. If several extraneous noises corrupt the measurement of interest, several adaptive filters can be used in parallel as long as suitable noise reference signals are available within the system.

Adaptive noise cancelling has been used for several applications. One of the first was a medical application that enabled the electroencephalogram (EEG) of the fetal heartbeat of an unborn child to be cleanly extracted from the much-stronger interfering EEG of the maternal heartbeat signal.

10.5 Inverse Modeling

We now consider the general problem of *inverse modeling*, as shown in Figure 10.8. In this diagram, a *source signal* $s(n)$ is fed into an unknown system that produces the input signal $x(n)$ for the adaptive filter. The output of the adaptive filter is subtracted from a desired response signal that is a delayed version of the source signal, such that

$$d(n) = s(n - \Delta) \, ; \qquad (10.12)$$

where Δ is a positive integer value. The goal of the adaptive filter is to adjust its characteristics such that the output signal is an accurate representation of the delayed source signal.

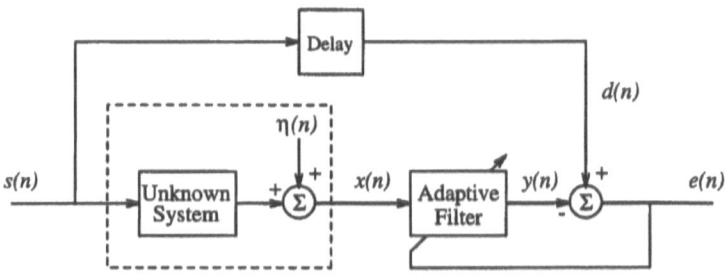

Figure 10.8. Inverse modeling.

The inverse modeling task characterizes several adaptive filtering applications, two of which are now described.

10.5.1 Channel Equalization

Channel equalization is an alternative to the technique of channel identification described previously for the decoding of transmitted signals across nonideal communication channels. In both cases, the transmitter sends a sequence $s(n)$ that is known to both the transmitter and receiver. However, in equalization, the received signal is used as the input signal $x(n)$ to an adaptive filter, which adjusts its characteristics so that its output closely matches a delayed version $s(n - \Delta)$ of the known transmitted signal. After a suitable adaptation period, the coefficients of the system either are fixed and used to decode future transmitted messages or are adapted using a crude estimate of the desired response signal that is computed from $y(n)$. This latter mode of operation is known as *decision-directed adaptation*.

Channel equalization was one of the first applications of adaptive filters. Today, it remains as one of the most popular uses of an adaptive filter. Practically every computer telephone modem transmitting at rates of 9600 *baud* (bits per second) or greater contains an adaptive equalizer. Adaptive equalization is also useful for wireless communication systems. A related problem to equalization is *deconvolution,* a problem that appears in the context of geophysical exploration. Equalization is closely related to *linear prediction,* a topic that we shall discuss shortly.

10.5.2 Inverse Plant Modeling

In many control tasks, the frequency and phase characteristics of the plant hamper the convergence behavior and stability of the control system. We can use a system of the form in Figure 10.8 to compensate for the nonideal characteristics of the plant and as a method for adaptive control. In this case, the signal $s(n)$ is sent at the output of the controller, and the signal $x(n)$ is the signal measured at the output of the plant. The coefficients of the adaptive filter are then adjusted so that the cascade of the plant and adaptive filter can be nearly represented by the pure delay z^{-1}.

18.5.3 Linear Prediction

A third type of adaptive filtering task is shown in Figure 10.9. In this system, the input signal $x(n)$ is derived from the desired response signal as

$$x(n) = d(n - \Delta) \tag{18.13}$$

where Δ is an integer value of delay. In effect, the input signal serves as the desired response signal, and for this reason, it is always available. In such cases, the linear adaptive filter attempts to predict future values of the input signal using past samples, giving rise to the name *linear prediction* for this task.

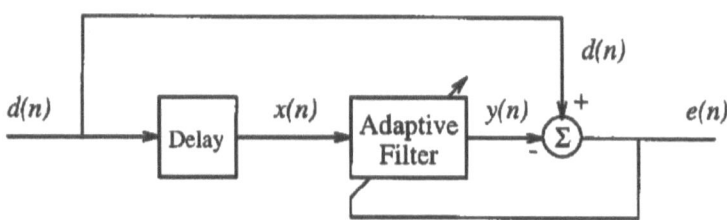

Figure 10.9. Linear prediction

If an estimate of the signal $x(n + \Delta)$ at time n is desired, a copy of the adaptive filter whose input is the current sample $x(n)$ can be employed to compute this quantity. However, linear prediction has several uses besides the obvious application of forecasting future events, as described in the following two applications.

10.5.3.1 Linear Predictive Coding

When transmitting digitized versions of real-world signals such as speech or images, the temporal correlation of the signals is a form of redundancy that can be exploited to code the waveform in a smaller number of bits than are needed for its original representation. In these cases, a linear predictor can be used to model the signal correlations for a short block of data in such a way as to reduce the number of bits needed to represent the signal waveform. Then, essential information about the signal model is transmitted along with the coefficients of the adaptive filter for the given data block. Once received, the signal is synthesized using the filter coefficients and the additional signal information provided for the given block of data.

When applied to speech signals, this method of signal encoding enables the transmission of understandable speech at only 2.4 kb/s, although the reconstructed speech has a distinctly synthetic quality.

Predictive coding can be combined with a quantizer to enable higher-quality speech encoding at higher data rates using an *adaptive differential pulse-code modulation* (ADPCM) scheme. In both of these methods, the lattice filter structure plays an important role because of the way in which it parameterizes the physical nature of the vocal tract.

10.5.4 Adaptive Line Enhancement

In some situations, the desired response signal $d(n)$ consists of a sum of a broadband signal and a nearly periodic signal, and it is desired to separate these two signals without specific knowledge about the signals (such as the fundamental frequency of the periodic component).

In these situations, an adaptive filter configured as in Figure 10.9 can be used. For this application, the delay Δ is chosen to be large enough such that the broadband component in x(n) is uncorrelated with the broadband component in $x(n - \Delta)$. In this case, the broadband signal cannot be removed by the adaptive filter through its operation, and it remains in the error signal $e(n)$ after a suitable period of adaptation. The adaptive filter's output $y(n)$ converges to the narrowband component, which is easily predicted given past samples. The name line enhancement arises because periodic signals are characterized by lines in their frequency spectra, and these spectral lines are enhanced at the output of the adaptive filter.

10.6 Adaptive Noise Reduction

In many applications, for example at the receiver of a telecommunication system, there is no access to the instantaneous value of the contaminating noise, and only the noisy signal is available. In such cases the noise cannot be cancelled out, but it may be reduced, in an average sense, using the statistics of the signal and the noise process.

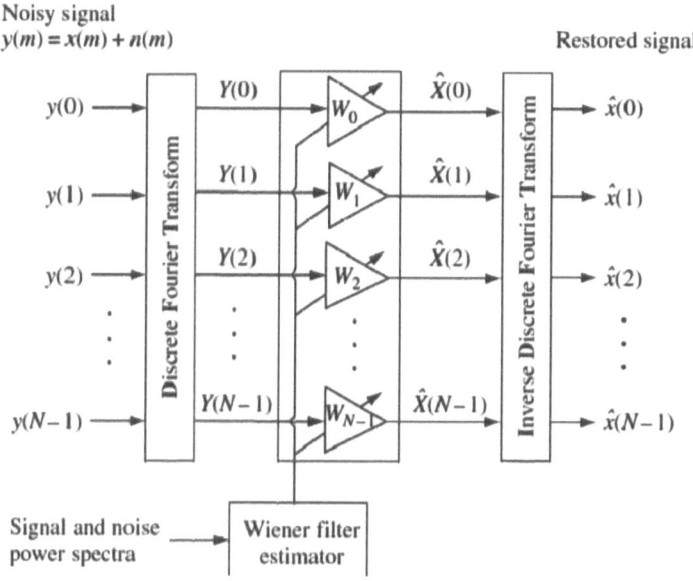

Figure 10.10. A frequency–domain Wiener filter for reducing additive noise

Figure 10.9 shows a bank of Wiener filters for reducing additive noise when only the noisy signal is available. The filter bank coefficients attenuate each noisy signal frequency in inverse proportion to the signal-to-noise ratio at that frequency. The Wiener filter bank coefficients are calculated from estimates of the power spectra of the signal and the noise processes.

Problems

10.1 Suppose the desired signal is given as:

$$d[n] = \cos[2\pi n/T_0].$$

The reference signal $x[n]$ that is applied to the adaptive filter input is given as

$$x[n] = \sin[2\pi n/T_0] + 0.5\cos[2\pi n/T_1],$$

where $T_0 = 5$ and $T_1 = 3$. Compute for a second-order system:

(a) $\mathbf{R}_{xx}$, $\mathbf{r}_{dx}$, and $\mathbf{R}_{dd}[0]$.
(b) The optimal Wiener filter weight.
(c) The error for the optimal filter weight.
(d) The eigenvalues and the eigenvalue ratio.
(e) Repeat (a)–(d) for a third-order system.

10.2. Suppose the desired signal is given as:

$$d[n] = \cos[2\pi n/T_0] + n[n],$$

where $n[n]$ is a white Gaussion noise with variance 1. The reference signal $x[n]$ that is applied to the adaptive filter input is given as

$$x[n] = \sin[2\pi n/T_0],$$

where $T_0 = 5$. Compute for a second-order system:

(a) $\mathbf{R}_{xx}$, $\mathbf{r}_{dx}$, and $\mathbf{R}_{dd}[0]$.
(b) The optimal Wiener filter weight.
(c) The error for the optimal filter weight.

(d) The eigenvalues and the eigenvalue ratio.
(e) Repeat (a)–(d) for a third-order system.

10.3. Suppose the desired signal is given as:

$$d[n] = cos[4\pi n/T_0]$$

where $n[n]$ is a white Gaussian noise with variance 1. The reference signal $x[n]$, which is applied to the adaptive filter input, is given as

$$x[n] = sin[2\pi n/T_0] - cos[4\pi n/T_0],$$

with $T_0 = 5$. Compute for a second-order system:

(a) **Rxx**, **rdx**. and **Rdd**[0].
(b) The optimal Wiener filter weight.
(c) The error for the optimal filter weight.
(d) The eigenvalues and the eigenvalue ratio.
(e) Repeat (a)–(d) for a third-order system.

10.4. Consider the process u(n) whose correlation function is $r_u(k) = a^{|k|}$ |. Determine the optimum coefficients of the one-step linear forward predictor of length two. Why is the second coefficient zero?

10.5. Consider the AR process

$$u(n) = 0.75u(n-1) + v(n),$$

and a process y(n) defined as

$$y(n) = u(n) + e(n) + 0.5e(n-1).$$

Both v(n) and e(n) are white noise with variance 1 and mean 0. Furthermore, they are independent. Determine the Wiener filter with the structure

$$\hat{u}(n) = w_0 y(n) + w_1 y(n-1).$$

10.6. Consider the Wiener filter problem with the observed signal u(n), the desired signal d(n). Set p = 1 in the method of steepest descent, where the weight vector is a scalar w(n).

(a) Determine the function J(n).
(b) Determine the Wiener solution wo(n) and the minimum estimation error Jmin.
(a) Sketch J(n) as a function of w(n).

10.7. Consider a so called adaptive line enhancer with a three-tap FIR filter and the following assumptions:

$$d(n) = \cos(\omega_0 nT + \theta) + v(n),$$
$$u(n) = d(n + \Delta),$$
$\theta \in R[-\pi, \pi]$ (uniformly distributed),

where $\omega_0 = 2\pi/3$ and $\Delta = 4$. Determine the optimal weights $\mathbf{w_0}$, and the minimum MSE Jmin for $\sigma_v^2 = 0.1$ and $\sigma_v^2 = 0.01$.

10.8. Determine the transfer function of the system shown below with the input signal d(n) and the output signal e(n), where the adaptive FIR filter has p coefficients. The signal (interference) u(n) is given by

$$u(n) = \cos(\omega 0 n + \varphi).$$

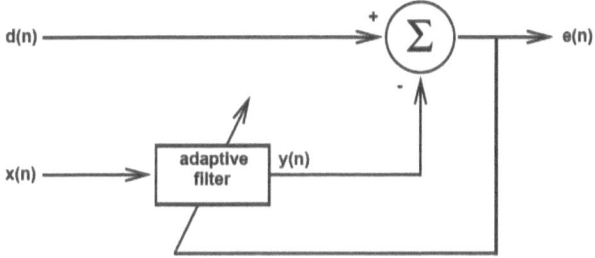

(a) Start with the relation between e(n) and y(n), i.e., $G(z) = Y(z)/E(z)$ Express $H(z) = E(z)/D(z)$ using $G(z)$.

(b) Write $u(n)$ as complex exponential function using Euler's formula.

(c) Now we concentrate on the scalar elements of the LMS update equation $w(n+1) = w(n) + \mu u(n)e(n)$, i.e.,

$$w_k(n+1) = w_k(n) + \mu u(n-k)e(n), \quad k = 0,\ldots,p-1. \quad (10.8)$$

(d) Express the z-transform of $u(n-k)e(n)$ using $E(z)$, the z-transform of e(n). Hint: use the relation $Z\{u(n)\} = U(z) \Rightarrow Z\{u(n)e^{j!0n}\} = U(ze^{-j!0})$ (where $Z\{\cdot\}$ is the z-transform).

(e) Determine the z-transform of (10.8) and express $W_k(z)$ as a function of $E(z)$.

(f) Determine the z-transform $y(n) = w^T(n)u(n) = \sum_{k=0}^{p-1} w_k(n)u(n-k)$ and apply the result from d), which yields $Y(z)$ as a function of $E(z)$

(g) The transfer function $G(z) = Y(z)/E(z)$ consists of two components where one can be neglected if p is large enough. Determine the resulting $G(z)$.

References

Ahmed, N., and Natarajan, T., Discrete-Time Signals and Systems. Reston, VA: Reston Publishing Co, 1983.

Akansu, A. N., and Haddad, R. A., Multiresolution Signal Decomposition: Transforms, Subbands, and Wavelets. Boston: Academic Press, 1992.

Alkin, O., Digital Signal Processing: A Laboratory Approach Using PC-DSP. Englewood. Cliffs, NJ: Prentice Hall, 1993.

Ambardar, A., Analog and Digital Signal Processing, Brooks/Cole, Monterey, 1999.

Antoniou, A., *Digital Signal Processing.* McGraw-Hill, New York, 2006.

Blahut, R. E., *Fast Algorithms for Signal Processing.* Cambridge University Press, New York, 2010

Box, G., Jenkins, G. and Reinsel, G., *Time Series Analysis: Forecasting and Control.* Wiley, New York, 2008.

Bracewell, R. N., *The Fourier Transform and its Applications.* McGraw-Hill, New York, NY, 2nd edition, 2000.

Brigham, E., *Fast Fourier Transform and Its Applications.* Prentice Hall, Upper Saddle River, NJ, 1988.

Burrus, C. S., Gopinath, R. A. and Guo, H., *Introduction to Wavelets and Wavelet Transforms: A Primer.* Prentice Hall, Upper Saddle River, NJ, 1998.

Carlson, G., Signal and Linear System Analysis. Wiley, New york (1998)

Carr, J. J., and Brown, J. M., Introduction to Biomedical Equipment Technology, 4th ed. Upper Saddle River, NJ: Prentice Hall, 2001.

Grover, D., and Deller, J. R., Digital Signal Processing and the Microcontroller. Upper Saddle River, NJ: Prentice-Hall, 1998.

Haykin, S.: Adaptive Filter Theory. Prentice Hall, Englewood Cliffs, 2001.

Haykin, S., Veen, B.V.: Signals and Systems. Wiley, New York, 1999.

Ifeachor, E. C., and Jervis, B. W., Digital Signal Processing: A Practical Approach, 2nd ed. Upper Saddle River, NJ: Prentice Hall, 2002.

Kehtarnavaz, N., and Simsek, B., C6X-Based Digital Signal Processing. Upper Saddle River, NJ: Prentice Hall, 2000.

Krauss, T. P., Shure, L., and Little, J. N., Signal Processing TOOLBOX for Use with MATLAB. Natick, MA: The MathWorks, Inc., 1994.

Ludeman, L. C., Fundamentals of Digital Signal Processing, John Wiley & Sons, 2003.

Lynn, P. A., and Fuerst, W., Introductory Digital Signal Processing with Computer Applications, 2nd ed. Chichester and New York: John Wiley & Sons, 1999.

Manolakis, D. G. and Ingle, V. K., "Applied digital signal processing, theory and practice, Cambridge University Press, 2011.

Mitra, S. K., Digital Signal Processing "A – Computer Based Approach, Tata Mc Graw Hill 2nd Edition, 2003.

Mitra, S. K., Digital Signal Processing: A Computer Based Approach, 3rd edn. McGraw-Hill, New York, 2006.

Oppenheim, A.V., Schafer, R.W., Buck, J.R.: Discrete-Time Signal Processing, 2nd edn. Prentice Hall Inc., NJ, 1999.

Oppenheim A. V. and Schaffer R. W. Discrete time Signal Processing, Prentice Hall Signal Processing,3rd ed, 2009.

Porat, B., A Course in Digital Signal Processing. New York: John Wiley & Sons, 1997.

Proakis, J.G. Manolakis, D.G., Digital Signal Processing: Principles, Algorithms and Applications, Macmillan, New York, 1996.

Proakis, J.G. Manolakis, D.G, Digital Signal Processing Principles, Algorithms and Applications, Pearson Education, 4th Edition, 2007.

Princen, J., and Bradley, A. B., Analysis/synthesis filter bank design based on time domain aliasing cancellation. IEEE Transactions on Acoustics, Speech, and Signal Processing, ASSP 34 (5), 1986.

Rabiner, L. R., and Schafer, R. W., Digital Processing of Speech Signals. Englewood Cliffs, NJ: Prentice Hall, 1978.

Rabiner, L R. and Gold, B., Theory and Application of Digital Signal Processing, Prentice Hall.

Soliman, S. S., and Srinath, M. D., Continuous and Discrete Signals and Systems, 2nd ed. Upper Saddle River, NJ: Prentice Hall, 1998.

Stearns, S. D., and Hush, D. R., Digital Signal Analysis, 2nd ed. Englewood Cliffs, NJ: Prentice Hall, 1990.

Stearns, S. D., Digital Signal Processing with Examples in MATLAB. Boca Raton, FL: CRC Press LLC, 2003.

Stearns, S. D., and David, R. A., Signal Processing Algorithms in MATLAB. Upper Saddle River, NJ: Prentice Hall, 1996.

Tan, L., *Digital Signal Processing.* Elsevier publications, 2007.

Vaidyanathan, P.P.: Multirate Systems and Filter Banks. Prentice Hall, USA, 1993.

Van der Vegte, J., Fundamentals of Digital Signal Processing. Upper Saddle River, NJ: Prentice Hall, 2002.

Vetterli, M., and Kovacevic, J., Wavelets and Subband Coding. Englewood Cliffs, NJ: Prentice Hall, 1995.

Widrow, B., and Stearns, S., Adaptive Signal Processing. Upper Saddle River, NJ:Prentice Hall, 1985.

Williams, A.B., Taylor, F.J.: Electronic Filter Design Handbook: LC, Active, and Digital Filters, 2nd edn. McGraw-Hill, New York, 1988.

www.ingramcontent.com/pod-product-compliance
Lightning Source LLC
Chambersburg PA
CBHW020725180526
45163CB00001B/117